Mining in Eastern Cassia County, Idaho

by Idaho Bureau of Mines

with an introduction by Kerby Jackson

Introduction

It has been years since the United States Geological Survey released this important publication. First released in 1931, this important volume has now been out of print for this days and has been unavailable to the mining community since those days, with the exception of expensive original collector's copies and poorly produced digital editions.

It has often been said that *"gold is where you find it"*, but even beginning prospectors understand that their chances for finding something of value in the earth or in the streams of the Golden West are dramatically increased by going back to those places where gold and other minerals were once mined by our forerunners. Despite this, much of the contemporary information on local mining history that is currently available is mostly a result of mere local folklore and persistent rumors of major strikes, the details and facts of which, have long been distorted. Long gone are the old timers and with them, the days of first hand knowledge of the mines of the area and how they operated. Also long gone are most of their notes, their assay reports, their mine maps and personal scrapbooks, along with most of the surveys and reports that were performed for them by private and government geologists. Even published books such as this one are often retired to the local landfill or backyard burn pile by the descendents of those old timers and disappear at an alarming rate. Despite the fact that we live in the so-called "Information Age" where information is supposedly only the push of a button on a keyboard away, true insight into mining properties remains illusive and hard to come by, even to those of us who seek out this sort of information as if our lives depend upon it. Without this type of information readily available to the average independent miner, there is little hope that our metal mining industry will ever recover.

This important volume and others like it, are being presented in their entirety again, in the hope that the average prospector will no longer stumble through the overgrown hills and the tailing strewn creeks without being well informed enough to have a chance to succeed at his ventures.

Kerby Jackson
Josephine County, Oregon
October 2015

CONTENTS

CONTENTS

CONTENTS

CONTENTS

ILLUSTRATIONS

ABSTRACT

Cassia County lies in the extreme south central part of Idaho, bordering Utah and Nevada. In this report about 2,000 square miles of the eastern part of the county are described.

The area is almost wholly in the Great Basin section of the Basin Range physiographic province, but its northern part extends into the Snake River Plains section of the Columbia Plateau Province. It is therefore a region characterized by isolated ranges (in part partially dissected block mountains) rising above aggraded desert plains. Toward the north the ranges end by plunging beneath the Snake River Plains, consisting of a young lava plain or plateau.

Within the district there are five mountain ranges, South Mountain, Albion Range, Malta Range, Black Pine Range, and the Sublett Range. Most of them rise abruptly from vast alluvial plains and are in a youthful to mature stage of dissection. Several bear distinct fault scarps which produced their block pattern. The Albion Range is the highest and rises to about 10,451 feet A.T. or to more than 5,000 feet above the alluvial plain at its northwestern base. The others are from 1,000 to 2,500 feet lower. These ranges bear witness of an interesting physiographic history and carry the remains of old erosion surfaces which may be correlated with the Snowdrift, Gannett, Elk Valley, Dry Fork, and Blackfoot surfaces described by G. R. Mansfield in southeastern Idaho. In addition, the Albion Range bears evidence of two stages of Pleistocene glaciation, the early considerably older and far more extensive than the younger. The Albion Range has also the further distinction of containing the interesting "Cassia City of Rocks," a small area of weirdly carved monoliths, spires, pinnacles and castellated forms in granitic rock in the heart of the Range. Less than half the area is mountainous. The two major basins are known as the Raft River Valley and the Goose Creek Valley. The climate of the region is semi-arid, with an average rainfall of less than 11 inches.

The rocks of the region comprise sedimentary beds of Proterozoic, Cambrian, Mississippian, Pennsylvanian, Permian, Upper Miocene (?), and Quaternary age, granitic rocks probably intruded in late Cretaceous or early Eocene time, flows of quartz latite and locally basalt intercalated in and lying above the tuffs and other strata of Upper Miocene (?) age, and flows of Snake River basalt of Quaternary age.

The Proterozoic strata are especially interesting, for they comprise a thick series of marine sediments, formerly mainly clean sandstones along with some limestone and shale beds, now entirely recrystallized and changed to vitreous quartzite, micaceous quartzite, marble, mica schist, and sillimanite schist. This series of rocks is new to the state and is evidently of pre-Beltian age. To it is given the name, Harrison series. The strata fall naturally into three lithologic units with a combined thickness of more than 9,000 feet. The lowest unit is the thickest and is composed of more than 6,000 feet of massive, evenly bedded white, generally fine-grained, vitreous quartzite, with occasional somewhat micaceous layers and thin beds of schist, the middle division consists of marbles, white quartzites, and schists, and the upper division is made up of white vitreous quartzite much like the lower.

The Cambrian beds are quartzites and limestones, but these have a very restricted distribution. The Mississippian series is represented by the Brazer formation only, which here consists of more than 2,000 feet of massive limestones with considerable sandstone and carbonaceous shale. The Pennsylvanian consists of the Wells formation of about 3,000 feet of cherty limestones and sandstones, and the Permian of the Phosphoria formation, of which the Rex chert is a conspicuous member.

There is no evidence of further sedimentation until Upper Miocene (?) times, when beds of gravel, sand, clay, marl, and volcanic ash were deposited. In the upper part of this series are intercalated flows of lava, mainly latites, with lava also capping the series. This series is correlated with the Payette formation in southwestern Idaho and also with the Salt Lake formation in southeastern Idaho. Quaternary deposits are widespread and include the vast alluvial fan deposits of the aggraded desert basins, terrace or beach deposits of old Lake Bonneville, glacial deposits of two distinct stages, loess deposits, hillside wash, and recent alluvial deposits.

The intrusive granitic rocks occur only in the Albion Range and South Mountain, and have in the main the composition of granodiorite, except at the margins of the stocks and the small batholith where there is reversal of the usual zoning and instead of being more basic, the marginal zone has the composition of granite, much of it porphyritic and gneissic. Pegmatites are abundant about the margins of some parts of the stocks and batholith. Weathering of the granitic rock in the main batholith in the southern end of the Albion Range has produced the strange "City of Rocks." The process involved has been largely one of granular disintegration, the result of hydration coupled with case-hardening or surface induration and other factors.

The Upper Miocene (?) lavas have the composition of quartz latites and quartz-tridymite latites. Differences in the flows are physical rather than chemical. Most flows have a thick basal zone of black porphyritic glass (feldspar vitrophyre), a central zone of pinkish or bluish-gray porphyritic aphanitic rock, and a thin top zone of black glass similar to that at the base. Basalt, locally intercalated in the series, is an olivine-rich variety in which the alteration of olivine to iddingsite has given the rock a faint reddish or purplish cast. The younger Snake River basalt is also an olivine-bearing basalt.

The highly metamorphosed state of the Harrison series suggests Proterozoic deformation in the region prior even to the deposition of the Beltian strata, for in localities nearest Cassia County the Belt series shows no such metamorphism as the Harrison. The most notable epoch of deformation occurred, however, at the close of the Cretaceous period during the Laramide revolution when the Carboniferous and older strata were complexly folded by great tangential compressive stresses acting from a westerly direction, and then sliced by great low-angled overthrust faults similar in every way to the well known overthrusts in southeastern Idaho. As a consequence the strata have been thrown into northerly and northwesterly trending synclinoria and anticlinoria, mainly overturned to the east, and the older rocks have been brought over the younger by the great overthrusts. It is because of this thrusting that the Proterozoic strata have been brought to view in the Albion Range, where they now rest on Carboniferous strata. This fault has been given the name of the Albion overthrust, is similar in all respects to the Bannock overthrust in southeastern Idaho, and probably represents a westward extension of the same general faulting. Following the great compression relaxational stresses were set up and the over-deformed mass was restored to a condition of equilibrium by normal faulting. Late in the deformative period, which probably extended also into the early Tertiary, came intrusion of a granitic magma and the development of the Cassia batholith and its outliers.

Erosion followed the building of the Laramide mountain system and continued in Cassia County more or less uninterruptedly to the Miocene, by which time the surface had been reduced to a peneplain. Uplift near the middle of the Miocene or earlier, perhaps augmented by faulting, caused the erosion of deep, wide valleys, which were later filled with lacustrine and fluviatile deposits of Upper Miocene (?) age, and finally more completely buried by great showers of volcanic ash and flows of latitic lava. The further record, to rather late in the Pliocene, is not readily deciphered, but apparently uplift occurred near the end of the epoch and erosion again reduced the surface to peneplanation before its close.

The next most conspicuous epoch of deformation is that which followed the late Pliocene planation and took place early in the Pleistocene when the earlier strata and structures were broken by great normal faults of northerly trend so characteristic of the Great Basin country and which have set off several of the ranges into their present tilted block outline. The Basin Range structure was thus superposed on the earlier structures. Movement along the fault planes apparently recurred until within comparatively recent times, as is evidenced from the youthfulness of the fault scarps and the partial destruction of several Pleistocene erosion surfaces. In the eastern part of the district the late block or normal faulting is not so pronounced and there remains evidence of a Pliocene erosion surface (Gannett), as well as of three Pleistocene cycles of erosion (Elk Valley, Dry Fork, and Blackfoot), each inaugurated by regional uplift.

After the early Pleistocene faulting had ceased, or had largely ceased, there came deformation of a different kind—the Snake River downwarp. Fault block and other ranges alike end against the present Snake River Plains by plunging beneath its surface, reappearing again in somewhat changed form in the mountains on the north side of that great depression. The subsidence apparently reopened older fault planes and through these issued great floods of the Snake River basalt which partly filled the vast depressed basin.

Glaciation occurred in the Pleistocene, apparently before and after the inauguration of the Blackfoot cycle of erosion. During the earlier epoch much of the east slope of the Albion Range was covered by glaciers from its crest to its base. The basin between it and South Mountain was also filled and from it tongues extended completely across the south end of the Albion Range.

Ore was discovered in the district about 1880, but the production has been small. Mineral deposits occur in both the Albion Range and the Black Pine Range, the former in the Stokes district and the latter in the Black Pine district. This distribution is genetic as well as geographic, as shown by the character and age of the mineralization. Deposits in the Stokes district occur dominantly as fissure veins in the pre-Cambrian strata and belong to the quartz-galena group of mesothermal character. These are centered more or less closely about a stock of granitic rock, a small outlier of the Cassia batholith, and are obviously related genetically to the late Cretaceous or early Eocene period of magmatic activity. In addition to a dominant quartz-galena filling, the veins contain inconsequential amounts of chalcopyrite, tetrahedrite, sphalerite, and pyrite. The galena is argentiferous. Some early production was obtained from the oxidized ores, but the enriched zone was shallow.

Unlike the deposits in the Stokes district, those in the Black Pine occur as irregular replacements of limestone and have as their chief metals silver and zinc. This group clearly belongs to the epithermal base-metal type and has for its essential minerals variable amounts of sphalerite, tetrahedrite (freibergite), jamesonite, and accessory pyrite, cinnabar and realgar, associated with a quartz, and locally barite and calcite, gangue. There are no igneous rocks in the vicinity of these deposits, but from the character of the deposits, including the nature of the mineralization and the similarity to epithermal base metal deposits elsewhere in the State and surrounding states, it is believed that the ores are genetically related to a younger Tertiary epoch of metallization, particularly to intrusive granites of late Miocene (?) age.

Other mineral resources in the district are building stones of various kinds, some of which have been utilized, limestone and marble, quartzite, volcanic ash, mica, feldspar, clay, cyanite, and road metal. Water resources and phosphate possibilities are not discussed herein. Another resource not yet utilized is the scenic attraction of the "Cassia City of Rocks." This unique area is surely worthy of development and deserves recognition as a National Monument.

Among the more notable contributions made as a result of the study of this area may be listed the following:

1. Recognition of a thick series of marine Proterozoic strata of pre-Beltian age which opens a new chapter in the geologic history of the State and which contributes additional data to the pre-Cambrian history of the West.

2. Discovery of great low-angle overthrust faults similar to those farther east in the Rocky Mountain province. Their occurrence serves to extend the typical Rocky Mountain structures much farther west than heretofore was supposed.

3. Recognition that the Basin Range faulting of the block or normal type is superposed on the Rocky Mountain structures and that this faulting is in no way related to the earlier epoch of deformation.

4. Determination that the Basin Range faulting within the area is mainly early Pleistocene. This suggests that perhaps much of the normal faulting within the Basin Range province is of similar age and therefore much younger than the mid-Tertiary age usually assigned.

5. Confirmation of the Snake River Plains area as a great synclinal depression or downwarp whose basin has been partly filled with Snake River basalt. Age of the subsidence, based on the relation of the downwarp to involved Basin Range faults and to early Pleistocene erosion surfaces, is determined as mainly or probably wholly Pleistocene so far as Cassia County is concerned.

6. Correlation of the Tertiary sedimentary and volcanic strata with the Upper Miocene Payette formation in the southwestern part of the State and also with the Salt Lake formation in the southeast part, which in the past has been tentatively assigned to the Pliocene. This correlation, based on stratigraphic succession and tracing of the formation from one area to the other, makes the Payette and Salt Lake equivalent formations and both of Upper Miocene age.

7. Recognition of the erosion surfaces so admirably interpreted and described by G. R. Mansfield in southeastern Idaho as extending also into the south part of the State.

8. Recognition of two stages of Pleistocene glaciation within the district, the earlier much older and far more extensive than the younger (late Wisconsin).

9. Discovery that granitic intrusion followed the overthrusting ascribed to the Laramide revolution, which adds to the likelihood that the Idaho batholith, of which the Cassia batholith is considered an outlier, is also late Cretaceous or early Eocene and not late Jurassic or early Cretaceous, as has been recently suggested.

10. Establishment of two epochs of metallization within the district, the older characterized by mesothermal quartz-galena veins and genetically related to the late Cretaceous or early Tertiary magmas; the younger characterized by epithermal base metal replacements of limestone with sphalerite, tetrahedrite, jamesonite, cinnabar, and realgar, quartz, calcite and barite as the distinguishing minerals and supposedly related to Tertiary intrusives of later age not exposed at the surface.

11. Description of the "Cassia City of Rocks" and ascribing its origin to deep granular disintegration of granitic rock induced mainly by hydration and aided by conditions of aridity favorable to casehardening or surface induration of the disintegrated rock and by favorable structures within the granite, together with its physical setting in the heart of the Albion Range.

12. Discovery of the Phosphoria formation (Permian) which may prove to be of economic significance.

13. Recognition of changes in the character of the Brazer formation (Upper Mississippian) which suggest increasing approach to the western shore of the Carboniferous sea.

14. Petrographic descriptions of the granitic rocks, the Tertiary quartz latites and basalt, and the Snake River basalt.

15. Description of the geography of a region never before described and the assignment of names to two major mountain ranges.

THE GEOLOGY AND MINERAL RESOURCES OF EASTERN CASSIA COUNTY, IDAHO

By Alfred L. Anderson

INTRODUCTION

PURPOSE AND SCOPE OF THE INVESTIGATION

As a result of revived interest in mining in Idaho during the past decade, attention was naturally directed to Cassia County as a possible new field for development. The region seemed to offer attractive inducements, for it was known that some lead, zinc, and silver ores had been shipped from the district since 1880 and that other mineral resources existed. As most of the County had never received geological investigation, little or no geologic information was available and this factor served to increase general curiosity and brought urgent requests for a survey of its resources. In response to these requests the writer was detailed by the Idaho Bureau of Mines and Geology to obtain and make available such information as pertains to its general geology and its mineral resources.

Because only one field season could be devoted to the investigation, the geologic mapping and study of nearly 2,000 square miles of virgin territory of necessity had to be of broad reconnaissance nature. Special consideration was given, however, to the mineralization in the Stokes mining district in the Albion Range and to the mineralization in the Black Pine district in the Black Pine Range. Special study was given the non-metallic resources, such as building stone, limestone, marble, and others. The widespread distribution of the Phosphoria formation, which in southeastern Idaho contains valuable beds of phosphate, was definitely recognized, but there was no time to study the formation in the detail that is desirable, and it was not ascertained whether the Phosphoria formation contains phosphatic shales in this region or not.

Inasmuch as most of Cassia County has been a large blank on the geologic map of the State, every effort was made to fill this gap. A thick series of pre-Cambrian strata was recognized and mapped with as much detail as possible, but time did not permit its separation into several distinctive lithologic units. Some Cambrian strata were also recognized, but these form only a very small part of the map area. Carboniferous sedimentary rocks are also widespread, and the Mississippian series (Brazer formation) was distinguished and mapped separately from the others, but it was not feasible to differentiate between the Pennsylvanian series (Wells formation) and the Permian series (Phosphoria formation) on the geologic map. Tertiary strata (Payette or Salt Lake formation) are widespread but are so intimately intercalated with and so extensively capped by latitic lavas that they cannot be adequately represented on the map. As nearly half of the area is concealed by Quaternary deposits, the mapping of an area of such size was greatly simplified. Intrusive granitic bodies, supposed outliers of the Idaho batholith, were also distinguished, as well as vast floods of younger acidic lavas whose relations to the Tertiary strata have already been mentioned. These lavas cover large areas, and, because of their general simple structure, were readily mapped. Only one other formation was distinguished—namely, flows of Snake River basalt.

Special studies were made of the geologic structure of the district. It was thought that much additional data might be obtained on the Rocky Mountain structure so typically represented in southeastern Idaho and also that its relation to the typical structure of the Great Basin country might be definitely established. Special effort therefore was made to discover evidence of the great low-angled overthrusts and the relation of these to the fault block ranges which were suspected to exist in the region. Several of such great overthrusts were found and likewise much evidence of faulting so characteristic of the Great Basin country. It was also found possible to date the Great Basin structures with greater precision than heretofore has been possible. In addition, further information was obtained on the Snake River downwarp and in particular its relation to the previous structures.

Physiographic features were given consideration and a complicated topographic history was deciphered. It was thought that study of this region might afford a better understanding of the physical history of the State, concerning which there has long been so much controversy. Special attention was given the "Cassia City of Rocks," a comparatively unknown spot of weirdly carved monoliths, spires, pinnacles, etc., whose forms were known to be sculptured in massive granite but whose origin was an unsolved problem.

Stratigraphic problems were anticipated and effort was made especially to correlate the supposed Tertiary strata with the Payette formation of southwest Idaho and with the Salt Lake formation in the southeastern part of the State. Attempt was then made to determine the stratigraphic relations of these two known formations to one another. The pre-Cambrian strata were studied, especially with the idea that the sedimentation here recorded might aid in a better understanding of Proterozoic paleogeography in the general region. Curiously, the pre-Cambrian strata proved to be pre-Beltian and were recognized as new to the State.

It is within the scope of this report to treat the essential features of the geology and mineralization and to make such interpretations and conclusions as the facts obtained seem to justify. The district holds a strategic position and the key to the solution of several obscure geologic problems.

FIELD WORK

Field work started in the district on June 15, 1930, and continued with only minor interruption because of inclement weather until August 27, 1930. Progress was rapid, for between June 20 and August 10 the writer was assisted by Mr. D. C. Livingston, who rendered valuable service in preparing a topographic sketch map of the district and in aiding with geological mapping. This left the writer free to devote his entire time to an investigation of the mineral resources, to the study and interpretation of physiographic, structural, and stratigraphic problems, and to geological mapping.

Lack of a suitable base map was a serious handicap to geological mapping and prevented any work other than reconnaissance. Forest maps with topography were available for part of the area and were most useful, but most of the district had only township plats made by the General Land Office and surveyed as early as 1872. Most of these were wholly unsatisfactory as a base for geologic work. Some of the township plats had been resurveyed during the past two decades and these were of considerable value. Preparation of a suitable base map for the geology was therefore necessary, and this duty kept Mr. Livingston busy for most of the summer. Unfortunately, geologic mapping had to be done before the completion of the base map and it was necessary to use any available map to record geology. Fortunately, timber maps of the forest divisions were especially accurate and these covered the most critical areas in the district.

Control for the topographic sketch map was from points established during the old Wheeler survey, but it was soon discovered that the latitude and longitude of these points had been incorrectly given and that the points were more than half a mile from their correct position. Most of the mapping was then done from section and township corners by intersection on desired points and their elevations were obtained by calculation from vertical angles or from barometric readings.

ACKNOWLEDGMENTS

The writer is especially indebted to Mr. D. C. Livingston for his services in the field. His topographic sketch map clearly outlines the major physiographic features and gives strong clue to much of the structure. Mr. Livingston was assisted in this work by Mr. Carroll Livingston. Acknowledgment is also due Mr. Leslie R. Vance, a graduate of the School of Mines, who assisted the writer in his studies of the geology and mineral resources.

Others in the district have also contributed to the work, and the writer especially wishes to convey his gratitude to Mr. W. J. Burridge, President of the Silver Hills Mining Company. Appreciation is also extended the forest supervisor of the Minidoka National Forest at Burley, Idaho, who so kindly furnished maps and directions to prospects and watering places.

The writer also wishes to express his gratitude to Dr. E. S. Bastin of the University of Chicago, who so kindly read the section of the report dealing with the economic geology; to Dr. R. T. Chamberlin, also of the University of Chicago, for his helpful suggestions pertaining to the section on structural geology; and to Dr. J. H. Bretz, of the University of Chicago, who gave his time to the reading of the sections describing the physiography of the region. The writer is also greatly indebted to Dr. G. F. Loughlin of the U.S. Geological Survey, who read the section dealing with the pre-Cambrian stratigraphy and who offered some very valuable suggestions.

PREVIOUS GEOLOGIC WORK

Previous information on the district is confined to a single brief report by Larsen,[1] who described the occurrence of cinnabar in the Black Pine district on the eastern slope of the Black Pine Range. He also mentioned the presence of nearby deposits containing zinc, silver, and copper, or silver, lead, and zinc.

Some work, mainly of a reconnaissance nature, has been done in adjoining areas, and this work affords some clue to the geology within

---

[1] Livingston, D. C., Tungsten, cinnabar, molybdenum, and tin deposits of Idaho: Univ. of Idaho School of Mines Bull. 2, Vol. 14, 1919, with a chapter by E. S. Larsen, U.S. Geological Survey, on the occurrence of cinnabar near Black Pine, Idaho, pp. 65-67.

the district.   Lignite beds[1] have been described in the Goose Creek
Basin in the western part of the County, and a survey was made of the
water resources[2] in the same region.   Reconnaissance studies have also
been made in the adjoining counties[3] on the east, and also in the Snake
River Plains area[4] on the north.   The adjoining region in Utah[5] has
also been briefly treated.

[1] Bowen, C. F., Coal and lignite in Boise and Cassia counties, Idaho: U.S. Geol. Survey Bull. 531-H, 1913.

[2] Piper, A. M., Geology and water resources of the Goose Creek Basin, Cassia County, Idaho: Idaho Bureau of Mines and Geology Bull. 6, 1923.

[3] Piper, A. M., Possibilities of petroleum in Power and Onedia Counties: Idaho Bureau of Mines and Geology Pamphlet 12, 1924.

[4] Russell, I. C., Geology and water resources of the Snake River Plains of Idaho: U.S. Geol. Survey Bull. 199, 1902.

[5] Butler, B. S., Loughlin, G. F., Heikes, V. C., and others, The ore deposits of Utah: U.S. Geol. Survey Prof. Paper III, 1920.

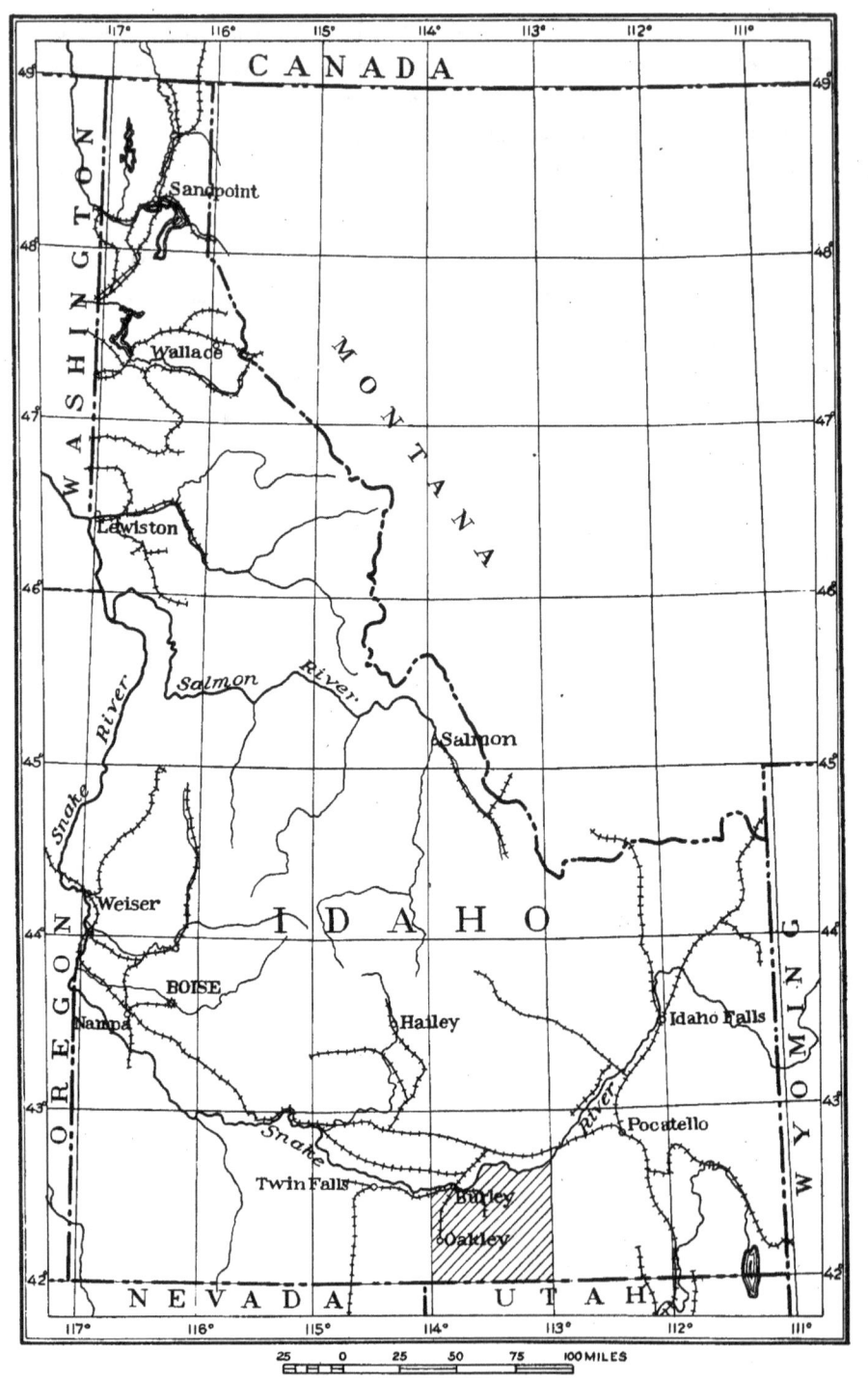

FIGURE 1.   Index map showing location of area.

GEOGRAPHY

LOCATION

The area under investigation lies wholly within Cassia County near the extreme south central part of Idaho bordering the State of Utah (Figure 1). It is bordered on the east by Power and Oneida counties, on the north by parts of Power, Blaine, and Minidoka counties. On the west the line separating R. 21 E. from R. 22 E., Boise meridian, was taken as the boundary. As outlined, the district lies mainly between 42° 00' and 42° 40' north latitude and between 113° 00' and 113° 50' west longitude.

TOPOGRAPHY

PHYSIOGRAPHIC SETTING

Eastern Cassia County lies mostly in the Great Basin section of the Basin and Range province[1] and is characterized by isolated ranges (in part partially dissected block mountains) rising above aggraded desert plains. Its northern part, however, merges abruptly with the Snake River Plains section of the Columbia Plateau province, whose characteristic feature is that of a young lava plain or plateau.

Most of the ranges are elongated in a northerly direction and those that reach the Snake River Plains section end by plunging beneath the lava-floored plateau. Most of them rise abruptly above vast alluvial plains and are in a youthful to mature stage of dissection. Several of the ranges still bear distinct physiographic evidence of the faults that produced their tilted block outlines. The larger intervening desert basins open to the Snake River Plains and are not sharply defined from it.

MOUNTAINS

There are five mountain ranges in the area, three of them of major magnitude. Of the five, two have never been officially named, and one of these is the largest range in the district, and perhaps the highest in Idaho south of the Snake River Plains. Three of the ranges lie near the west border of the district and two along the east side. Those on the west include South Mountain, Albion Range, and Malta Range, named in order from west to east. These are parallel, closely grouped ranges and are separated by only minor alluvium-floored basins. The names assigned to the Albion and Malta ranges are new. Those on the east

---

[1] Fenneman, N. M., Physiographic divisions of the United States: Annals of the Assn. of Amer. Geog., Vol. XVIII, No. 4, 1928.

side of the County are the Black Pine Range and the Sublett Range, named from south to north. Foothills, or the lower northern flank of the Raft River Range, extend into the southern part of the district. This range, unlike the others, is a domed uplift with its axial line at right angles to the elongation of the other ranges.

Most of the ranges exhibit notable differences among themselves, especially in the character of their dissection, determined mainly by the nature of the rocks which compose them. These differences can be presented only through individual description. Order of presentation is not based on the size of the ranges nor on their relative importance, but in their order from west to east as listed above.

SOUTH MOUNTAIN

South Mountain, also known as Middle Mountain, is a northerly extending prong of the Goose Creek Mountains of northwestern Utah and northeastern Nevada. It ends south of Oakley about 12 miles from the State line. This mountain has a pronounced tilted-block appearance with its crest or axis very near its eastern margin (see topographic sketch map). Its slope to the east is steep or abrupt and resembles a little-eroded fault scarp. Its slope to the west is longer and much more gradual, approximating an angle of about 18°. Streams on its western slope have carved relatively shallow valleys, streams on the east side are few and have caused little migration of the drainage divide. A comparatively low depression extends diagonally across the range to the northwest and divides it into two sections, one offset slightly from the other. The crest of the two sections is angular and attains an elevation above 8,000 feet A.T., or nearly 2,000 feet above the floor of the narrow basin which separates it from the Albion Range. These outstanding characters are well represented on the topographic map.

Several valleys have been carved in the steep eastward facing scarp and these have the characteristic appearance of glaciated valleys (Plate I, A). Such wide, shallow, U-shaped valleys are characteristic of only this range and the Albion and Raft River ranges. The basin which separates South Mountain from the Albion Range also has the appearance of a glaciated trough. Low, much-eroded patches of moraine at the mouths of many of the U-shaped valleys confirm the glacial origin of these valleys.

ALBION RANGE

The Albion Range is probably the most imposing range in Idaho south of the Snake River Plains. It is 36 miles long, extending into

Idaho from Utah, and ends at the north by plunging beneath the Snake River Plains. It rises to a maximum elevation of about 10,451 feet A.T. or more than 5,000 feet above the plains at its base. But in spite of its size and prominence it has never received an official name, having been mentioned only in connection with its two highest parts, Mount Harrison and Cache Peak. Because the U.S. Forest Service has designated this part of the Minidoka National Forest as the Albion Mountain division, it is here proposed that the term, Albion, be adopted to designate the entire range.

The Albion Range is sharply set off from the parallel ranges on each side, although not widely separated from either. It rises more or less steeply and on its western and northwestern sides presents a bold, even front, little indented by the steep valleys on its flanks. East of Oakley the range swings in a gradual curve to the northeast. Its eastern side is less regular and is interrupted by three basin-like embayments, named respectively from north to south the Albion, the Elba, and the Almo basins. At these embayments there is a very noticeable narrowing of the range, giving the range the appearance of two broad domes. The most southerly dome culminates in Cache Peak at 10,451 feet A.T. Its width there is about 10 miles. The broad dome to the north culminates in Mount Harrison at an elevation of about 9,200 feet A.T. Its distance across is about 12 miles.

The summit of the range is not sharp as in the other ranges in the district, but is broad and notably flat or level. This character is suggested from a distance by the evenness of the sky-line and is strikingly confirmed from the summit itself. Its surface is very gently rolling, as illustrated in Plate I, B, and is covered with a deep mantle rock. Its gentleness of surface is quite out of harmony with the steep flanks of the range and it has every appearance of an "old land" surface, uplifted and subsequently scarcely affected by erosion. This surface is exceptionally well retained about Mount Harrison, where it lies at an elevation between 8,000 and 8,500 feet A.T., or from 500 to 800 feet below the highest crest of the range. The highest crest itself carries another similar surface of about two square miles whose appearance suggests the remnant of a still more ancient surface (Plate II, A), but it is very likely a detached part of the other surface brought to its present elevation as a result of comparatively recent normal faulting. Even more of the main summit erosion surface remains about Cache Peak and, as shown in Plate II, B, forms a distinct broad terrace on both sides of the peak. This surface extends for nearly five miles south and west of the peak (or behind the peak as shown in the picture).

Cache Peak is apparently a monadnock on this surface. The erosion surface may correspond with the Snowdrift peneplain of pre-Middle Miocene age described by Mansfield[1] across some of the higher ridges in the mountains in southeastern Idaho. As will be explained later, the old erosion surface in the Albion Range may be an exhumed surface from which Upper Miocene strata have been stripped off by erosion.

Another striking character of the range is its notable lack of dissection except by shallow U-shaped valleys with uniformly steep gradients (Plate I, A, Plate II, B, and topographic map). These broad shallow valleys lie mainly on the east side of the range and especially on the slopes which descend to the three basins. Their form is strongly suggestive of glacial action, particularly as the valleys are remarkably straight and some of them, at their mouths, have minor surface irregularities suggestive of old moraines. Additional criteria of glacial origin are afforded near the south end of the range near Almo, where the U-shaped valleys extend directly across the range as illustrated in Plate III, A, and have their heads in the basin which lies between South Mountain and the Albion Range. These valleys are not occupied by streams except for what water is collected within the range. The topographic relations strongly suggest that the basin between the two ranges was once a center for ice accumulation from which glaciers moved eastward across the Albion Range and northward up the basin and over the low divide into Birch Creek drainage. The glacial stamp in this part of the range is especially well shown on the topographic map.

In a few places the characteristic smooth slopes of the range have been sculptured into sharper lines and the higher parts given a more alpine appearance. This work is obviously the product of a younger stage of glaciation, much less extensive than the one that carved the long U-shaped valleys. The development of high cirques is the most outstanding feature of this glaciation. Mount Harrison bears such a cirque on its northeast side, as illustrated in Plate II, A, and in the floor of the cirque nestles beautiful Lake Cleveland (Plate III, B). The floor of the cirque is not much below the level of the old erosion surface. A cirque on the southeast side of Harrison is much deeper but much less perfect. Cache Peak has also been sculptured by mountain glaciers. On its north side is a large cirque in which lie five small lakes, known locally as Independence Lakes. These are at an elevation of 8,500 feet A.T. Cache Peak has also been much steepened on its southeast side where the descent is nearly vertical for 3,000 feet.

---

[1] Mansfield, G. R., Geography, Geology, and Mineral Resources of part of Southeastern Idaho: U.S Geol. Survey Prof. Paper 152, 1927, p. 14.

Glaciers of this epoch did not form at levels below 8,000 feet A.T., nor did they extend much below that elevation.

MALTA RANGE

The Malta Range is a young tilted block mountain at the east base of the Albion Range. It is about 34 miles long and ends on the south about three miles from the Utah line and on the north about two miles beyond the end of the Albion Range, where it, too, terminates by plunging beneath the Snake River Plains. It attains an elevation of about 8,200 feet A.T. in its highest part opposite Cache Peak, more than 3,000 feet above the desert plain at its east base. This range is as wide as the Albion Range and nearly as imposing from its eastern side. It has not received a name, in spite of its prominence. It is here proposed to name it the Malta Range from the town not far from its eastern base.

This range is very similar to the young fault block ranges in southern Oregon, and, like them, has its crest very near the margin of the block, with a steep, precipitous scarp facing the east (Plate XIV, B) and a long gradual slope to the west (see topographic sketch map). Its steep eastern scarp has not been greatly modified by erosion where the range is low, but along higher parts it has been much broken by landslides. Some of the landslide blocks have traveled several miles from the base of the range. The back or west slope of the range shows the same progressive dissection from early youth in its lower northern end to early maturity in its higher parts farther south. Its tilt-block appearance has been mainly retained because of the late date at which the range was formed and because it bears a comparatively resistant capping of lava above a base of very easily eroded tuffs. Wherever streams have cut through the lava into the tuff, destruction of the block form has been rapid, hastened largely by landslides. The west slope of the range blocks the Albion, Elba, and Almo basins, and, between them, the Malta Range actually lies against the lower slope of the Albion Range. Two streams, Cassia Creek and Raft River, have maintained their courses across the faulted and tilted block, thereby dividing the range into three sections, separated by narrow canyons.

Glaciers have left a mark on the high middle section of the range where outlines of cirques are faintly discernible. These glaciers extended to somewhat lower altitudes on the east side of the range than on the west, but not far below the crest on either side. The glaciated valleys with their headward cirques are mostly cut through the lavas into the underlying soft tuffaceous beds, and this has greatly favored the development of landslides.   This land movement has largely destroyed the earlier glacial forms.

BLACK PINE RANGE

The Black Pine Range is in the southeast corner of Cassia County and, like the Malta Range, does not extend outside the district.   It is isolated from all other ranges and rises steeply from broad piedmont alluvial slopes and irregular pediments or bordering rock floors to a maximum elevation of 9,700 feet A.T. near its central part, or 3,800 feet or more above its base.   From a distance it appears as if composed of three segmented masses as shown in Plate IV, A, with the two end segments plunging in opposite directions to the level of the surrounding alluvial plain.   Its southern end lies approximately at the Utah line and its other end about 17 miles to the north.   The range is wider than most in proportion to its length, having a maximum width of nine miles.   Its plan is not elliptical, as its domical profile might suggest, but it is mainly bounded by straight lines.   Its north end appears as though it had been sliced obliquely and this feature destroys what would otherwise have been a rectangularly blocked range (see topographic sketch map).

Unlike the Albion Range, the Black Pine has been deeply dissected into a stage of early maturity, and the valleys, instead of being wide and shallow, are narrow and steep, as shown in Plate IV, B.   Summits comprise narrow ridges of irregular profile, bounded by exceedingly steep slopes.   Dissection has not been simple, but the erosional cycle has been several times interrupted by uplift during which youthful valleys have been carved in the floors of the older valleys.   The present valleys are extremely youthful and have steep sides several hundred feet high.   It is probable that this young stage of dissection corresponds to the late Blackfoot cycle, described by Mansfield,[1] in the mountains of southeastern Idaho.   Above the present stream valleys lie the remnants of older broad and shallow valleys in the Black Pine Canyon region in the southern end of the range, these more than a mile wide.   This earlier stage of erosion may correspond to the Dry Fork Cycle in the mountains of southeastern Idaho, also described by Mansfield.   Above this surface the slopes rise very steeply, but still older surfaces are not sufficiently well preserved to admit their interpretation and correlation.

Features of glacial origin near the summit of the range have not been well preserved, for the floors of the high glacial troughs have been

---

[1] Prof. Paper 152, op. cit., p. 17.

largely destroyed by subsequent valley cutting. From the position and character of outwash on the northeast flank of the range it is probable that two epochs of glacial erosion are recorded, the earlier at an elevation of 500 feet and the younger 200 feet above the present valley bottoms.

Shore features of ancient Lake Bonneville are well displayed along the lower eastern and southeastern slope of the range. These constitute a series of two or three excellently preserved beach terraces. It is interesting to note that, had Lake Bonneville not found outlet elsewhere, a 50-foot increase in the lake level would have caused it to spill over Kelton Pass at the south end of the range into Raft River Valley.

SUBLETT RANGE

About half of the Sublett Range is in northeastern Cassia County and the remainder in Power and Oneida counties. It, too, is an isolated range, much longer and broader than the Black Pine, and rises steeply, except on the north, above aggraded desert plains and minor bordering rock floors or pediments, but with slopes and borders less regular than in the other ranges. It ends on the north by plunging gently beneath the Snake River Plains. By reason of its gradual descent several of its higher parts have been detached or isolated from the main range and rise as islands along its northwestern border. Summit levels are lower than in nearby ranges and the highest seldom exceed 7,500 feet A.T. The mountain, therefore, loses the range-like aspect of its neighbors and appears more as a broad and maturely dissected low upland or plateau. Its pattern is complex and consists largely of parallel ridges, trending northwestward, and locally northeastward, separated by steep narrow valleys. Drainage lines and ridge patterns are obviously greatly influenced by the character and structural trends of the underlying strata.

This plateau-like area has been deeply dissected and everywhere has steep slopes. Its erosional history is complicated but much more easily interpreted than the stages recorded in the development of the Black Pine Range. At least four cycles of erosion are recorded on the ridge tops and in the present valleys (Plate V). These probably are to be correlated with the Gannett, Elk Valley, Dry Fork, and Blackfoot cycles interpreted and described by Mansfield[1] in the mountains farther east. The earliest represents a widespread erosion surface in late stages of maturity or early old age now preserved on the narrow crests of the

1 Prof. Paper 152, op. cit., pp. 12-19.

highest ridges in the range. Below are ridge crests that record the floors of broad and relatively shallow valleys, produced probably during the Elk Valley cycle. Again, below the terraces and uplands of the Elk Valley surface and above the present stream valleys, lies a group of high-level valleys or terraces ascribed to the Dry Fork cycle. This surface lies from 300 to 500 feet below the next highest surface and in most places about 300 feet above the bottoms of the present stream valleys. The rejuvenation of the region by uplift at the close of the Dry Fork cycle enabled the streams to cut their present sharp-featured canyons in their valley floors. These valleys are narrow, with sides so steep that landslides are conspicuous features and in parts of some valleys, as along upper Lake Fork, have caused serious obstruction to drainage. None of the surfaces have been affected by glaciation and it seems probable that the range, because of its relatively low elevation, escaped glaciation almost entirely.

RAFT RIVER RANGE

Raft River Range is mainly outside the district in Utah, but as part of its northern slope lies within the south margin of the district, and as its structural features aid in solving some of the structural problems within Cassia County, it will be given brief description. This range differs from the others in that it is a partially dissected anticlinal uplift or domed range with its axis trending from east to west at right angles to the elongation of the other ranges. Its eastern end is south of Black Pine Range and is separated from it by the low, wide Kelton Pass. Its west end appears to merge with the Grouse Creek or Goose Creek Mountains. Most of its northern flank appears as a dip slope of moderate steepness, considerably modified by erosion. Its summit appears notably broad and flat and attains an elevation greater than the highest point in the Black Pine Range. General surface features are much the same as those in the Albion Range and South Mountain and the slopes have many broad shallow valleys and ridges with broad gently rounded crests. Piedmont alluvial slopes extend some distance up its flanks.

BASINS AND PLAINS

RAFT RIVER VALLEY

Raft River Valley forms a vast aggraded alluvial plain from 10 to 15 miles wide, enclosed by the Black Pine and Sublett ranges on the east, Raft River Range on the south, Malta Range on the west, and open to the Snake River Plains on the north. Its origin is entirely structural and not erosional, but its floor has been raised to its present level by the waste from the flanking ranges and in its northern part by flows of Snake River basalt. In places at the base of the Black Pine and Sublett ranges the alluvial plain joins narrow irregular rock floors or pediments. From its lower or central parts the alluvial plain rises at first almost imperceptibly and then more steeply far up the flanks of the ranges. In places a difference in elevation of more than 1,000 feet is shown between the edge and the center of the basin. The intervening slopes represent broad alluvial fans, which coalesce into a broad piedmont alluvial plain. The appearance of this wide valley is well pictured in Plate IV, A, taken near the south end of the valley, across to the Black Pine Range 11 miles distant.

In three places the Raft River Valley is separated from other basins by low divides, little higher than its own margin. One of these reentrant basins lies between the Black Pine Range and the Sublett and joins with the broad basin east of the Black Pine Range. Another low basin or reentrant lies between the Black Pine Range and the Raft River Range. This low, wide divide is known as Kelton Pass, and descends gradually on the other side to Curlew Valley north of Salt Lake. Another lies south of the Malta Range, but differs from the others in that it is not so much a divide as a pass from a higher level basin (Almo Basin) to a lower level basin.

GOOSE CREEK VALLEY

Goose Creek as it enters Idaho from Utah and Nevada flows through a narrow valley or canyon, but at the reservoir about three miles south of Oakley the valley opens into a wide aggraded plain that spreads fanlike to the north and eventually merges with the Snake River Plains not far south of Burley. At Oakley the valley plain is about five miles wide, and where it probably merges with the Snake River Plains it has a width of 12 to 15 miles. Its surface is more level than that of Raft River Valley, mainly because its origin is more largely due to obstructed drainage induced by flows or floods of Snake River basalt. Bordering piedmont alluvial slopes are much less conspicuous than about the Raft River Valley.

SNAKE RIVER PLAINS

The Snake River Plains is the dominant physical feature of southern Idaho. It is essentially a vast young lava plain or plateau, which stretches as an arc concave to the north across the south part of the State from Wyoming to Oregon, a distance of 400 miles, and which forms a surface from 50 to nearly 125 miles wide. As will be discussed in greater detail in another section, the lava plain represents a partial filling of a great downwarp or geosyncline, called by Kirkham[1] the Snake River Downwarp. Its surface is generally featureless, except for occasional low lava domes. This plain crosses the northern end of the district, and, as mentioned earlier, covers the downwarped ends of the Albion, Malta, and Sublett ranges. The boundary separating the Plain from the Raft River and the Goose Creek basins is difficult to define and must be arbitrary. A line connecting the ends of the ranges probably is as good as any other that might be selected.

DRAINAGE

All the drainage, except that from the east and southeast slope of the Black Pine Range, is northward and eventually runs into the Snake River either in surface streams or underground. Raft River and Goose Creek are the two main tributaries of Snake River in this part of the State, but at present only Raft River carries surface water the entire distance to the Snake. Waters in Goose Creek have been entirely diverted for irrigation in the Goose Creek Valley.

Raft River pursues an interesting course through the district. One of its main headward branches, Junction Creek, has its source in the basin between South Mountain and the Albion Range, where, joined by Cottonwood Creek, it flows southward along the basin and enters the South Fork of Raft River about four miles from the State line in Utah. From there Raft River swings to the east and crosses the Albion Range, and, on the other side, turns to the northeast into Almo Basin. It is there joined by several tributaries from the east slope of the Albion Range and also several from the north slope of the Raft River Range. Here it again turns to the east and flows through a narrow valley across the southern end of the Malta Range into the main Raft River Valley. Its course is then Northward to Snake River, and, for most of the distance, it is nearer the western side of that great valley than the eastern. Several miles northeast of Idahome, however, it has been deflected around the edge of a Snake River basalt flow to the east margin of the wide valley and continues along the edge of the flow to the Snake River. Near the center of Raft River Valley it is joined by Cassia Creek, one of its principal tributaries, and in the upper end of the Valley by Clear Creek. The latter drains the north slope of the Raft River Range, whereas Cassia Creek with its numerous headward

[1] Kirkham, V. R. D., A geologic reconnaissance of Clark and Jefferson and parts of Butte, Custer, Fremont, Lemhi, and Madison counties, Idaho: Idaho Bureau of Mines and Geology Pamphlet 19, 1927, pp. 6-14, 24-25; also Snake River Downwarp, Jour. Geology, Vol. 39, 1931, pp. 456-482.

tributaries rises on the slope of the Albion Range bordering Elba Basin and then flows directly across the middle part of the Malta Range to Raft River Valley.

Goose Creek has most of its tributaries outside of the district. Birch Creek, also rising in the basin between South Mountain and the Albion Range, flows northwestward and joins Goose Creek not far from Oakley. It has numerous tributaries from the west slope of the Albion Range as well as several smaller ones from the east slope of South Mountain. Although streams are relatively abundant along the west slope of the Albion Range to the north, all of them soon disappear at the edge of the alluvial plain before reaching Goose Creek. Curiously enough, Goose Creek is also turned to the northeast in the northern part of Goose Creek Valley by a flow of Snake River basalt and its former channel joins the Snake River at Burley.

The Albion Range is well watered and most of its valleys carry permanent streams at least to the base of the mountain. Several of the larger ones already mentioned actually have sufficient volume to carry them to Raft River or Goose Creek. It is not feasible to mention all the streams, but Marsh Creek and its tributaries, which supply water for irrigation in Albion Basin, are worthy of description. This system drains the northeast slope of the Albion Range, flows northeast through the center of Albion Basin, and then, unlike Raft River or Cassia Creek, does not continue across the Malta Range but turns northward along the west base of the range to the Snake River Plains. On reaching the Snake River Plains it swings sharply to the west along the outer edge of the Snake River basalt and enters the Snake River west of Declo.

Malta Range has no permanent streams and few springs. The Black Pine Range is similar. The latter is thoroughly dissected by canyons, but only Eight Mile Canyon, Six Mile Canyon, and Kelsaw Canyon on the west side of the range carry water, and only Sweetzer Canyon on the northeast side. None of these streams extend far from the base of the range. Some of the other canyons with intermittent streams have names, although dry for most of the year. Most notable of these are Rice Canyon, Formation Canyon, and Black Pine Canyon on the south end of the range, and Mineral Gulch, East Dry Canyon, and Pole Canyon on the east side.

Drainage of the Sublett Range is mainly by intermittent streams. Sublett Creek, with its tributary, Lake Fork, is the largest stream in the range and carries a large volume of water throughout the year. South Heglar Canyon, North Heglar Canyon, and Calder Creek represent the main drainage systems in the north end of the range, but these carry no water at present except at their very heads.

CASSIA CITY OF ROCKS

Erosional forms of extraordinary interest have been carved in granite in the southern part of the Albion Range west of Almo. These are enclosed in a large basin excavated in the core of the range and are entirely hidden from view on the west by the high quartzite flank of the range and partly hidden from the east by a row of hogback ridges.

Within the basin the rock city is unfolded in all its silent splendor. It is the site of a maze of weirdly carved forms scattered aimlessly about the basin, but more closely grouped near its upper western margin. The forms are distributed over an area of about nine square miles or a distance of nearly six miles from north to south. Their distribution is not uniform, but resembles scattered villages or hamlets with more widely scattered outlying forms between. One of the larger villages lies in the upper drainage of Johnson and Almo creeks, but the one with most features of interest and at the same time most easily accessible, lies in the upper drainage of Circle Creek along the road between Almo and Oakley. Entry into the rock city is either from Oakley over the high western rim of the basin, from Almo through valleys across the hogback ridge, or from the south end of the basin.

Such features as found herein are probably no better displayed anywhere in the world, and probably surpass the curious Buffalo rocks[1] in the Buffalo Mountains in Australia. These groups of strange and bizarre rocks are among the most attractive features of the region and inspire a feeling of awe whenever seen. It is not easy to describe them. No two groups or individuals are alike—there is diversity in the individuals and in the character of the groups. The illustrations must be closely examined and studied to appreciate the details in these curious forms. General views of the groups and some of the individual members are afforded in the illustrations (Plate VI and following). Some of the individuals appear as great domed monoliths rising high above the floor of the basin (Plate VI, B). Others rise to lesser elevations and offer greater detail and variety of forms. Near the southern end a row of glistening turrets and fortresses stretch across a low saddle keeping guard of the road (Plate VII, A) and the city behind. Elsewhere rocks project as towers and spires from 100 to 150 feet above the basin floor. Some of the forms resemble miniature skyscrapers or bear marked

[1] Dunn, E. J., The Buffalo Mountains: Victoria Geol. Survey Memoir No. 6, 1908.

resemblance to some of the interesting rock walls in the famous Zion Canyon in Utah (Plate VII, B). Others resemble curious Oriental mosques or temples (Plate VIII, A). Taken as a whole, one is reminded of some strange fairy land.

Closer observation leads to the discovery of even more curious and fantastic forms. Not the least of these are the groups of rocks and individuals that have quaint resemblance to human forms or images and to animals. Some of these appear as natural as though sculptured by human hands. Most curious of these is the "Old Woman" carved in the granite ledge with her side to the wall, as so strikingly brought out in Plate VIII, B. It requires little more imagination to see other forms on the same rock, much as in an ancient Grecian frieze. Other forms have quaint resemblances to elephants, birds, dragons, chickens, toadstools (Plate IX, A), and have earned such names as "Old Hen with Her Chicks," "Dragons Head," "Elephant Rock," "Giant Toadstool," and others. In addition there are pedestal rocks (Plate IX, B), natural bridges, caves, bathtub rocks (Plate XIII, B), hollow boulders (Plate XIII, A), and other weird and grotesque forms (Plate XI and Plate XII). Additional features and forms will be pointed out later in the section where the origin of this fantastic city is discussed.

The rock city is easily accessible and not far from the main line of travel (Old Oregon Trail, U.S. Highway No. 30), yet it is scarcely known to the people within the surrounding region and entirely unknown to the world in general. It has vast possibilities as a national playground and is surely worthy of consideration as a National Monument.

CLIMATE

The aridity of the region is at once suggested from its inclusion within the Basin and Range physiographic province. The mean annual precipitation in the southern part of the County is about 10 inches and in the northern part 11.58 inches. In places favored by higher relief the total rises somewhat, but in parts of Raft River Valley the precipitation is much less. Much of the precipitation comes in the winter months in the form of snow and in the spring months as rain. Summer and autumn months are usually cloudless, but local convectional storms accompanied by torrential downpours are not at all uncommon, especially in midsummer. During 14 years the Minidoka Irrigation Project in the northwestern part of the County has been favored on an average with 238.6 clear days, 57.8 partly cloudy days, and 68.8 cloudy days per year. Without irrigation the successful growing of crops generally has proved impossible.

Because of its altitude and aridity the daily range in temperature is great at all seasons. During the summers the days are hot, but, owing to the clear atmosphere, radiation is rapid and the nights are cool, a daily range from 100°F. at mid-day to 50°F. at night being not uncommon. During the winter the cold is moderately severe and varies considerably with the altitude. Temperatures as low as —16°F. have been recorded at Oakley and 0°F. is not unusual anywhere in the district for short periods. Even during the period of severe cold in winter, daily temperature ranges of 30°F. are not uncommon. Killing frosts rarely occur between May 16 and October 15 in the areas adapted to farming.

VEGETATION

The vegetation adapts itself closely to the climatic conditions, varying in character and quantity with exposure to sun and wind and with the distribution of moisture and temperature. The moister bottoms are covered with reeds, rushes, and wild grasses. The lower mountain slopes and foothills together with the exposed upper slopes and drier bottoms and broad basins are largely covered with sage brush, interspersed with grasses and weeds that are utilized for grazing.

Arid though the region is, timber is not wholly absent. The lower foothills are generally studed with juniper, which, though rarely exceeding 15 feet in height, is valuable as a source of fence posts. On favorable slopes of the upland where the precipitation exceeds the average, there are thick stands of lodge pole pine, Douglas fir, spruce, and possibly other conifers. Such patches of timber are seen mostly on the high north slopes of the ranges and only rarely on the high eastern slopes. From the south the Black Pine Range appears absolutely barren of timber, but from the north many small patches are seen on the steep slopes. Forested slopes covering several square miles occur on parts of the Albion Range, but generally the patches are much smaller and end against the sage brush with great abruptness. The Sublett has more timber than the others and has supplied several small sawmills. Poles are removed yearly from all the forested areas. Some protected basins and creek bottoms are choked with dense aspen growths. Some of the rocky places in all higher parts of the district are covered with mountain mahogany.

CULTURE

Burley, the county seat, with a population of approximately 5,000, is the largest town. It is in the heart of the Minidoka Irrigation Project and is the trade center for a large surrounding area.   Principal produce of the Project are potatoes, sugar beets, red clover seed, alfalfa, beans, grain, poultry, swine, sheep, and dairy cattle.

Albion, in the heart of Albion Basin, is a small agricultural community, whose principal asset is the State Normal School.   Oakley is also in the center of an irrigated tract in the Goose Creek Valley.   Its population is nearly 1,300, to which may be added about 800 more for the population of the entire tract.   Its agricultural products are essentially the same as those of the Minidoka Project.   Elba and Almo, each in their respective basins on the east side of the Albion Range, are in small irrigated tracts.   Idahome in Raft River Valley is the terminus of a branch railroad and its chief asset is a grain elevator.   Malta, a small community on Raft River bottoms, is a trade center for the middle Raft River Valley.   Strevell, Naf, and Stanrod are small communities near the Utah line.   Naf and Stanrod are in small farming tracts at the base of the Raft River Range.   Strevell has a wireless station and landing field belonging to the air mail service, and in addition, a hotel.

Dry farming is practiced on the borders of the Sublett Range with notable success along its west margin near Sublett post office and also near the northwest part of the range in the youthful to submaturely dissected, loess-covered high basin trenched by the lower Heglar canyons.   Wheat is the only produce of the dry farm district.

The mountainous regions are practically uninhabited and serve as grazing grounds or range lands for large flocks of sheep and large herds of cattle.   The Sublett and Black Pine ranges are especially given over to the grazing of cattle, and it is probable that cattle are nearly as numerous in the Albion Range.   Grazing of sheep and cattle is one of the most important industries in the district.

TRANSPORTATION

The Minidoka-Buhl branch of the Oregon Short Line (Union Pacific system), which crosses the northwestern part of the district, supplies the principal outlet for the region.   From Burley a branch line extends up the Goose Creek Valley to Oakley.   A second branch also extends from Burley to Idahome in the lower end of the Raft River Valley. All other towns, except Declo, are without direct rail connection. Declo is on the Idahome branch.

An excellent system of county, state, and national roads makes most parts of the district readily accessible.   The Old Oregon Trail, now the main arterial highway across southern Idaho, extends entirely across the northern end of the district and passes through Declo and Burley and the heart of the south side of the Minidoka Irrigation Project. About three and one-half miles from the west margin of Raft River Valley it is joined by the South or Ogden branch of the Old Oregon Trail, which enters the Raft River Valley over Kelton Pass and extends its full length, passing through Strevell, Malta, and Idahome. Oakley is joined to Burley, 27 miles distant, by a graded and graveled highway. From Oakley a secondary road in fair condition leads southeastward along the old Boise-Kelton stage route up Birch Creek and across the Cassia City of Rocks. Another road extends directly east and across the Albion Range and descends to Elba. An excellent surfaced highway also joins Albion with the Old Oregon Trail near Declo. From Albion a secondary road passes southward through Elba and Almo to the Cassia City of Rocks. Secondary roads also extend from it to Raft River Valley along Cassia Creek and also through Raft River Narrows east of Almo. The highway from Malta to Sublett warrants special mention, as well as its extension into Black Pine Basin. The basins and foothills of the ranges are nearly everywhere accessible, mostly on roads which are not maintained and seldom used except by sheep wagons. The Heglar region is served by roads deep with mud at certain seasons of the year and as deep with dust at other seasons. Roads in fair condition extend far up the valleys in the Sublett Range, but few do so in the Black Pine and Albion Ranges.

GENERAL GEOLOGY

STRATIGRAPHIC GEOLOGY

GENERAL OUTLINE

Stratified rocks are widely distributed in Cassia County, but the stratigraphic column is far from complete. The Proterozoic era is represented by a great thickness of sedimentary marine strata of probable pre-Beltian age; the Paleozoic by Cambrian (?) and Carboniferous sediments; and the Cenozoic by sedimentary and volcanic strata of mid-Tertiary age and by unconsolidated Quaternary deposits.

Time allowed for study in the district did not permit detailed studies of the stratigraphy and accurate measurements of sections. Most of the formations differ little in character and thickness from their respective occurrences in nearby areas where they have been studied in greater detail.

The stratigraphy is briefly summarized below:

STRATIGRAPHY OF EASTERN CASSIA COUNTY

Thickness (feet)

QUATERNARY:
 Recent: Alluvium.
 Pleistocene: Hillwash and older alluvium; moraine and glacial outwash; and loess. Undifferentiated from Recent alluvium on the geologic map.
Unconformity.

TERTIARY:
 Upper Miocene (?): Payette or Salt Lake formation (chiefly stratified ash, tuff, conglomerate, clay, and marl, and in its upper part interbedded with latite, rhyolite, and basalt flows) .... 2,500±
Unconformity.

CARBONIFEROUS:
 Permian: Phosphoria formation (shale, limestone, and chert, possibly with phosphatic shales) .... 700±
 Unconformity.
 Pennsylvanian: Wells formation (chiefly cherty limestones and sandstones) .... 2,900±
 Unconformity.
 Mississippian: Brazer formation (massive limestone, subordinate carbonaceous shale and sandstone) .... 2,000±
Unconformity.

CAMBRIAN (?): Undifferentiated quartzite and limestone .... 800±
Unconformity.

PRE-CAMBRIAN (?): Harrison series (chiefly quartzite, lesser schist and marble) .... 9,000+

PRE-CAMBRIAN ROCKS

Strata of pre-Cambrian age are widely distributed in the Albion Range and South Mountain, and, outside of the district, in the Raft River Range. These comprise a thick series of quartzites and a lesser thickness of schist and marble. Neither the base nor the top of the series is exposed, and its exact thickness is undetermined, although more than 9,000 feet of beds occur in the district. The most complete section is across the northern part of the Albion Range, where the strata have been arched into a broad anticline. As the best exposures are on the flanks of Mount Harrison, it is herein proposed that the series be known as the Harrison series.

HARRISON SERIES

CHARACTER AND DISTRIBUTION

The Harrison series falls naturally into three well defined lithologic units, a lower division composed mainly of quartzite, a middle division of quartzite but containing some schist and marble, and an upper division of quartzite. Of the three divisions the lowest is by far the thickest. It was not feasible to separate these on the geologic map in the time available for study, especially as structural doming, faulting, and inadequate exposures make tracing of beds exceedingly arduous. The shallowness of the valleys on the slopes of the ranges also has not been favorable for exposing adequate sections. All three divisions are represented in the northern section of the Albion Range, but only two, and largely the lower, appear in the southern half.

Lower division: The lower division of the Harrison series is most widespread along the east flank of the Albion Range from the south edge of the Albion Basin to the Utah line. It forms there the exposed east limb of a broad anticline or dome, but as the dome has not been symmetrically eroded and as the axis lies near the east margin of the range, the lower division disappears beneath the younger members westward. It does, however, appear on the west flank of the range south of Cache Peak, where the anticlinal axis lies near the center of the range.

The lowest part of the division is mainly a light gray to white pure quartzite with thick, even bedding, but it contains some slightly micaceous members. As much as 3,000 feet of these quartzites appear in the hogback ridges which enclose the Cassia City of Rocks. Very little of this part, however, appears in the northern section of the range, for there erosion has not carved as deeply into the series. Above, there appears about 2,000 feet more of slightly micaceous quartzites alternating with numerous thick beds of pure quartzite and a few thin beds of mica schist. This part is best displayed in the northern section of the range, especially in Conner Creek Valley, and on the north slope facing Albion Basin. Most of it has been eroded from or is concealed in the Cache Peak section, except on its western side. The upper part of the division is mostly a thick, even bedded quartzite, mainly light gray or white, much like the lower part. It has some darker micaceous beds. It is distinguished readily from the lower part, however, by the presence of a thin pebble bed, whose character has been changed under the intense metamorphic agencies to a conglomerate gneiss. Pebbles have been much squeezed and rudely orientated as in an augen gneiss, and are in part outlined by tiny foils of greenish muscovite. Microscopic examination reveals complete recrystallization of the original constituents and the development of greenish muscovite, and microcline. Zircon and apatite are accessories. This bed is conspicuous on Howell Creek, Conner Creek, and on the slope east of Cache Peak. Thickness of the entire upper part is probably not less than 1,500 feet.

Middle division: The middle division is most widely distributed in the north section of the Albion Range, mainly on its west side, but also on its crest and over nearly its entire part north of Mount Harrison. It extends along the west flank of the southern section only as far south as Carson Creek. From there to the Utah line it has been entirely removed by erosion. Likewise only a small patch remains at the north end of South Mountain.

This division differs markedly from that below in containing much more schist and also two or more marble or limestone members. Its maximum thickness is probably no less than 2,000-3,000 feet. The boundary between this and the lower division cannot as yet be accurately defined, but it lies probably not far beneath the lowest calcareous member.

The lowest part of the division is well exposed on the ridge that encircles Lake Cleveland. Here the series is composed of more than 500 feet of impure crystalline limestone or marble, calcareous quartzite, pure quartzite and schist, the calcareous beds comprising several hundred feet of the whole. The marble is, in part, micaceous and has small amounts of phlogopite. Above, the division becomes increasingly micaceous and arenaceous and is mainly a micaceous quartzite with thin beds of garnetiferous mica schist. Near the center of the division the rock is mainly massive white quartzite, rather evenly bedded, but with a few slightly micaceous beds and thin beds of schist. This zone is probably not more than 1,500 feet thick. Its character is much the same as that in the central part of the lower division. These beds form the crest of Mount Harrison and are widespread to the north. They are in turn overlain by a second series of crystalline limestones or marbles, schist, and thin beds of quartzite, the entire series measuring probably more than 500 feet in thickness but not more than 1,000 feet. The exact number of marble beds has not been satisfactorily determined. In places but one bed measuring from 60 to 100 feet was seen, in others there seem to be several beds. Where but one marble bed appears, other members have probably been removed by faulting. Much of the marble is massive, pure white or mottled, but some is thinly bedded. In one place a bed of massive white and mottled bluish-gray marble is followed by a much thicker series of thinly bedded partly calcareous gneiss, interbedded with layers of medium to coarse grained, indistinctly banded marble, several hundred feet thick, and again overlain by a bed of grayish-white laminated marble. The entire thickness of the calcareous series is perhaps over 400 feet. Much of the white marble is a high-calcium carbonate, but the bluish-gray facies is somewhat dolomitic. A thin section of a more intensely altered facies of the marble shows in addition to coarse calcite grains, some muscovite or talc, quartz, titanite, apatite, and magnetite, also some large grains of hematite. Evidently, the marble has in part been subjected to contact metamorphism as well as to dynamometamorphism.

A schist member at the top of the marble series is especially worthy of description. This member is about 150 feet thick, and may be traced without much interruption along the wide saddle between the two main sections of the Albion Range, as well as along the west flank of the north section and over much of the north end of the range. It shows little change in character from place to place, but consists essentially of a very fine grained grayish-black schistose rock studded with small crystals of pinkish garnet. Microscopic study is necessary to identify other minerals because of their minute size. These consist of variable amounts of quartz, graphite, sericite, chlorite, biotite, andalusite, sillimanite, cyanite, plagioclase, microcline, magnetite, titanite, rutile, apatite, and zircon. These are not uniformly distributed through the member, but give rise to layers richer in one or more of the constituents, usually in quartz, graphite, and sillimanite. Some of the lower layers are more quartzose and are comparatively rich in chlorite and have few other minerals. In other parts sillimanite needles predominate and with graphite form the main constituents. Andalusite invariably accompanies the sillimanite, usually in much subordinate amount, but in places it exceeds the sillimanite. Its usual occurrence is in small grains or rods or as much larger metacrysts crowded with other mineral constituents. Cyanite is only abundant locally, especially in one locality near the north end of the range where it forms nests or masses of long-bladed crystals in the garnetiferous schist. These lenses of cyanite attain a foot in thickness and the individual blades a length of seven inches. Other minerals are always very minor accessories, except garnet, which is uniformly distributed through the schist in excellent crystals seldom more than one-eighth of an inch in diameter and generally closely spaced. Most of the schist might be classed as a garnetiferous sillimanite schist. The presence of such alumina-rich minerals as sillimanite, andalusite, and cyanite strongly implies that the original rock was a clay shale.

Upper division: The upper division of the Harrison consists of white dense vitreous quartzite, in part massive with ill-defined bedding and in part distinctly bedded or layered in beds six to eighteen inches thick. Exposures occur at the base of the Albion Range east of Oakley near Slide Creek and may compose the main quartzitic rock in South Mountain. It is also exposed along the base of the north section of the Albion Range in beds much disturbed by faulting. From 400 to 500 feet of these beds may be seen where the Albion highway crosses the range south of Declo, but the exact thickness of the section cannot as yet be given, as higher parts have been eroded.

CORRELATION AND AGE

The Harrison series is continuous with the ancient strata in the Raft River Range which comprise the greater part of that range, and which have been assigned to the Algonkian.[1]

Rocks most similar to the Harrison series are those in the Hailey quadrangle[2] on the north side of the Snake River Plains about 90 miles distant. These have been divided into two formations—a lower series of arenaceous strata 6,600 feet thick, the Hyndman formation, much the same as the lower division of the Harrison; and an upper series of quartzose and calcareous strata at least 1,560 feet thick, the East Fork formation, much like the middle division of the Harrison. Similarities are sufficiently striking to suggest correlation of the two groups of rocks. Those in the Wood River region have been assigned tentatively to the Algonkian largely because of their great degree of metamorphism, much

[1] Butler, B. S., Loughlin, G. F., Heikes, V. C., The Ore deposits of Utah: U.S. Geol. Survey Prof. Paper 111, 1920, p. 78.

[2] Umpleby, J. B., Westgate, L. G., and Ross, C. P., Geology and ore deposits of the Wood River region, Idaho: U.S. Geol. Survey Bull. 814, 1930, pp. 9-17.

greater than that shown in adjoining masses of Paleozoic rocks. Some of the metamorphism is assigned to igneous emanations, but it is noted that the contrast in metamorphism between this group and the Paleozoic is marked whenever the formation boundaries are crossed, irrespective of distance from intrusive rocks. On the other hand, C. P. Ross, who has worked in the Custer quadrangle, immediately northwest of the Hailey quadrangle, reports limestone with some quartzite similar in lithology and degree and character of metamorphism to the East Fork formation, apparently stratigraphically well above dolomite which contains Silurian fossils. The weight of present evidence in the Hailey quadrangle, however, favors the conclusion that the Hyndman and East Fork formations are probably Algonkian.

The Harrison series is surely not to be correlated with the Belt series of late Algonkian age so widely distributed in northern Idaho and western Montana, nor is the series to be correlated with the late Algonkian rocks in the Wasatch Range in Utah. Not only does the Harrison series show a much greater degree of metamorphism than the members of the Beltian series and other rocks of supposed Algonkian age, but it also shows an entirely different manner of origin, being made up mainly of marine sediments instead of continental deposits as in the other series. The Belt rocks are characterized by their general lack of metamorphism, except in the vicinity of granitic intrusives, and are composed in large part of sandstones, shales and argillites, and lesser limestones, deposited mainly in shallow waters and distinguished by sun cracks and ripple marks, cross-bedding, and highly colored layers, especially reddish or purplish and greenish. The rocks in the Wasatch Range are more or less similar and are described by Blackwelder[1] as alternating beds of sandstone or quartzite, slate, and conglomerate, not greatly indurated, composed of poorly assorted materials, with sandy beds, yellow, gray, and red, and shaly layers, purple, maroon, and green, further distinguished by much cross-bedding, ripple marks, and mud cracks. Beneath this series he has found another, however, composed of mica schist and metaquartzites.[2] This characterization of the Belt series and its supposed equivalent in Utah is vastly different from the thick white quartzite series with its beds of schist and marble in Cassia County and the Raft River Range in Utah, where the rocks are especially distinguished by their whiteness of color, well assorted materials and regularity of bedding, and by their extreme degree of metamorphism, which has been sufficient to change the calcareous

---

[1] Blackwelder, Eliot, New light on the geology of the Wasatch Mountains: Geol. Soc. Amer. Bull., Vol. 21, 1910, pp. 517-542.
[2] Blackwelder, Eliot, Wasatch Mountains revisited: Geol. Soc. Amer. Bull., Vol. 36, I 25 abstract, pp. 132-33.

members to coarse marble, the shale to sillimanite schists with accessory andalusite, cyanite, and garnet, and the included pebble beds to conglomerate gneiss. The great degree of metamorphism is regional and is not due to igneous intrusion, for the series shows the same degree of metamorphism throughout, only slightly enhanced near the contact with the intrusive granite bodies. As additional evidence that the great metamorphism is not the result of igneous intrusion, may be cited the lack of metamorphism in the Carboniferous limestones and sandstones lying beneath the overthrust block of the metamorphic series east of Oakley in near proximity to the granitic stocks.

Local and regional evidence therefore favors an ancient date for the meta-sedimentary series, probably pre-Beltian. It is likely that the Harrison series is the equivalent of the new series of pre-Cambrian rocks mentioned by Blackwelder[1] in the Wasatch Mountains, which consist largely of silvery mica schists and white metaquartzites, whose age is intermediate between the gneisses of supposed Archean age and the quartzite-slate series in the Cottonwood canyons, which he regards as probably late Algonkian. The Harrison series is thus very likely older than the Belt series or the late Algonkian rocks and perhaps belongs in the central or lower part of the Proterozoic, perhaps to the Huronian, if such a correlation can ever be made between the western pre-Cambrian and that of the Lake Superior region or the Canadian shield.

CAMBRIAN SYSTEM

Strata belonging to the Cambrian are believed to lie along the north base of the Raft River Range as windows beneath an over-riding sole of the Harrison series. As only a narrow border extends across the State line into Idaho there was little opportunity afforded for studying this system.

These beds appear to be made up of a lower series of massive quartzites, in part fine grained and in part conglomeratic, white to pinkish in color, more or less similar to the Brigham quartzite in southeastern Idaho described by Mansfield.[2] The quartzite has been so broken by major and minor faults that little could be learned of its lithologic sequence or of its thickness, except that it exceeds 300 or 400 feet. Above it lies a series of shales and grayish limestones, probably corresponding to the Middle Cambrian limestones in southeastern Idaho. But here again the limestones have been so sheared because of their

[1] Geol. Soc. Amer. Bull., Vol. 36, 1925, op. cit., p. 132.
[2] Prof. Paper 152, op. cit., p. 52.

nearness to the overlying fault zone, that they have a distinct schistosity, sufficiently severe to obliterate all fossils that may have existed. More than 500 feet of the grayish limestone, usually with poorly preserved bedding, appear on the slope west of Stanrod. The quartzite occurs mainly at the base of the range between Stanrod and Naf.

Cambrian rocks are widely distributed over western Utah[1] and everywhere the early or lower strata are composed of quartzite and sandy shale and the upper part of limestones and dolomites. Much variation in thickness occurs, but in the west and north part of the State the Cambrian attains a thickness of more than 9,000 feet.

CARBONIFEROUS SYSTEM

Strata of Carbonifeous age are well represented in eastern Cassia County, where they constitute the only rocks in the Black Pine Range, most of the rocks in the Sublett Range, and two small areas in the Albion Range. All three series of the Carboniferous system are present, the Mississippian, Pennsylvanian, and Permian. These consist chiefly of limestone and sandstone, together with subordinate cherty and shaly members, including possible beds of phosphatic shale, and differ little in lithology and thickness from the Carboniferous rocks described in southeastern Idaho.

MISSISSIPPIAN SERIES

Mississippian strata include only the Brazer formation, which differs from that farther east in containing a little more shale and sandstone. Its presence is largely inferred from its lithologic resemblance to the Brazer limestone elsewhere. The Madison limestone, which lies below the Brazer formation, was not recognized nor was its presence suggested from faunal studies. As the base of the Brazer was nowhere definitely established, it is not likely that erosion has brought the Madison limestone to view.

BRAZER FORMATION

The Brazer formation is believed to occupy most of the southern segment of the Black Pine Range, the northern end of the Sublett Range, and the small patch of Carboniferous rocks at the north end of the Albion Range. In the Black Pine and Sublett ranges it appears as an overthrust block on younger Carboniferous strata, but in the Albion Range it appears as a window beneath an over-riding block of pre-

---

[1] Prof. Paper III, op. cit., pp. 78-79.

Cambrian strata. In all places its presence is mainly inferred from the massiveness of some of its limestone members and their tendency to form pronounced ledges.

In southeast Idaho, according to Mansfield,[1] the Brazer is characterized by massive, generally light gray limestone that is prominent as a ridge and cliff marker and which in certain beds is highly fossiliferous. Especially conspicuous and characteristic are cup or horn corals distinguished by their large size and numerous slender septa. Locally the limestones are arenaceous and interbedded with sandstones. Strata similar to these occur in Cassia County, but a thick shale bed is also included.

Massive, white or light gray limestone is topographically conspicuous in the southern end of the Black Pine Range, where beds form cliffs and ledges and otherwise produce rugged country. The base was not definitely established, but the lowest part exposed consists of several hundred feet of black carbonaceous shale with a little sandstone above and below and interbedded with thin bedded dark gray limestone. This part of the series is well exposed at the heads of Black Pine and Kelsaw canyons. Thin bedded limestones, some members abundantly fossiliferous, alternating with some massive beds continue for several hundred feet and possibly a thousand feet above the shale horizon. Above this appears the most characteristic part, consisting of about 400 feet of non-fossiliferous light gray massive limestone, usually much broken and shattered but with the fractures healed with calcite. Bedding is generally indistinct and the limestone forms the most pronounced ledges in the district (Plate X, A). In part it is slightly cherty, but for the most part it is composed of relatively pure limestone adapted for industrial use. This member forms the host for most of the replacement mineral deposits in the Black Pine district and may be readily traced across the range by the cliffs and ledges it produces. Above are some more thin bedded dark gray limestones, non-ledge forming, with some intercalated argillaceous members and sandstones. Fossils collected from the series in Black Pine Canyon in Sec. 8, T. 16 S., R. 29 E., and identified by Mr. George H. Girty of the U.S. Geological Survey are as follows:

Triplophyllum ? sp.	Productus aff. inflatus
Stenopora sp.	Pugnoides sp.
Rhombopora sp.	Productus phosphaticus ?
Productus cora	Eumetria ? sp.

Mr. Girty refers the lot to the Brazer on one count only. He states that except for the forms listed as *Eumetria ? sp.*, this formation might

---

[1] Prof. Paper 152, op. cit., p. 63.

better be Wells than Brazer, and if that one form could be interpreted as some uncommon or unknown species of *Hustedia*, the lot might be assigned tentatively to the Wells. This collection is apparently from the limestones above the main light gray massive cliff-forming beds.

Strata in the north part of the Sublett Range differ little from those mentioned above. The lower part, where exposed in the lower walls of South Heglar Canyon, consists of black carbonaceous shales and some sandstone or quartzite overlain by nearly a thousand feet of thin and massively bedded limestones, some fossiliferous. In this horizon is also a bed of intraformational conglomerate or breccia about 50 feet thick, made up of sharp angular pieces of black and reddish chert. Above is about 60 feet of massive limestone, then thinner beds of more shaly character. In the steep canyon walls the limestones form frequent ledges (Plate X, B), but on more gentle slopes outcrops are usually not so conspicuous. Above the breccia and shaly limestones again occur the light gray massive ledges so characteristic of the Black Pine area and form the series of pronounced ledges which make up the upper part of Cedar Peak ridge and also the white ledges on Badger Peak to the north. Again no fossils were found in them. A collection made from ledges in this series of strata in Sec. 7, T. 11 S., R. 30 E., in rocks mapped as Pennsylvanian by Piper[1] in Power County contains the following fauna:

Lithostrotion sp.	Crinoid stems
Campophyllum sp.	Euomphalus ? sp.

In addition to these, large cup or horn corals are abundant. Mr. Girty states that the fauna of this lot suggests the Brazer fauna more strongly than any other.

Another collection from North Heglar Canyon in Sec. 10, T. 11 S., R. 30 E., in the limestones below the upper massive members contains:

Stenopora aff. carbonaria	Productus semireticulatus
Polypora sp.	Productus cora
Rhombopora sp.	Pustula aff. porrecta

Mr. Girty much more tentatively includes these among the Permian lots. It is possible that these are from a small window of Permian rocks beneath the over-riding block of Brazer, but the field relations strongly suggest that this horizon belongs in the Brazer. Fossils in the hills north of the main range in Sec. 3, T. 10 S., R. 28 E., consist of cup corals and colonial types, *Syringopora sp.*

Strata in the north end of the Albion Range apparently consist mainly of the light gray massive limestone, with some beds of darker

---

[1] Piper, A. M., Possibilities of petroleum in Power and Onedia Counties: Idaho Bureau of Mines and Geology Pamphlet 12, 1924.

gray color and some of shale. Some of the beds contain poorly preserved forms of cup corals. It is possible that this same horizon is also exposed in places in the window of Carboniferous rocks west of Elba.

PENNSYLVANIAN SERIES

Pennsylvanian strata consist of the Wells formation alone. This formation resembles the Brazer but contains much more sandstone and much chert, especially in the limestones. It assumes great prominence in the Black Pine and Sublett ranges, but unlike the Brazer does not form cliffs or ledges, but everywhere the slopes are smooth (Plate IV, B, and Plate V, A). Its characters differ little from the Wells in southeastern Idaho. In some places it is associated with the overlying Permian series and because this association is so intimate, on account of folding and faulting it was not possible to differentiate the two series in the time available for work.

WELLS FORMATION

The Wells formation is much more widely distributed than the Brazer. It comprises the middle and northern segments of the Black Pine Range as well as part of the southern. It, together with the Permian series, comprises much of the southern and western part of the Sublett Range. It also is present in the mid-part of the Albion Range between Elba and Oakley, in the window beneath the pre-Cambrian rocks.

In southeast Idaho the Wells formation has three variable but fairly distinct parts—a lower sandy and cherty limestone facies, a middle sandy facies, and an upper siliceous limestone. The lower beds are generally more sandy and cherty than the massive beds in the upper part of the Brazer formation. The middle sandstone member forms steep slopes strewn with debris, but makes few or inconspicuous ledges except in the steeper-walled valleys and then rarely. The upper siliceous limestone member forms prominent ledges or knobs along the hillsides, but in some places it is absent.

Cherty limestones and sandstones of the lower division are especially well displayed in the north section of the Black Pine Range. The limestones are dark gray to bluish gray and the chert is largely in the form of nodules but occurs also in bands and irregular masses or streaks. On weathering, the surface is usually strewn with the chert masses and fragments. These sandy and cherty limestones and their interbedded sandstones have an observed thickness of about 800 feet. Some horizons are fossiliferous with fossils both in the sandy and cherty limestone members. Fossils collected in Sec. 25, T. 14 S., R. 28 E., have been identified by Mr. Girty as follows:

Triticites secalius ?	Productus semireticulatus
Stenopora sp.	Spirifer triplicatus
Polypora sp.	Composita subtilitia
Rhombopora sev. sp.	Plagioglypta ? sp.
Derbya ? sp.	

These Mr. Girty definitely assigns to the Pennsylvanian (Wells formation).

Alternating beds of cherty and sandy limestone and sandstone also make up rocks in the southeast part of the Black Pine Range. These are similar to the strata in the north end of the range and although no fossils were obtained from them, probably belong to the lower division of the Wells. These are apparently overlain by Brazer beds as a result of thrust faulting.

It is possible that the middle segment of the Black Pine Range has some rocks of the lower division, but most of the strata belong to the middle and upper divisions. The middle sandy section is nearly 2,000 feet thick, and in addition to sandy limestones has much sandstone. It forms topographically the highest part of the Black Pine Range but fails to give good ledges (Plate IV, B), although the steep slopes are well strewn with debris. Most of the sandstone, both in debris and in the outcrop, has a characteristic brownish color, but fresh rock is gray. Above the sandstones on the steep west slope of the middle segment of the range is a light gray limestone member about 100 feet thick which tends to form ledges and which might represent the uppermost division of the Wells. Only poorly preserved fragments of fossils were found in it.

Both the middle and upper division of the Wells formation were recognized in the Sublett Range and it is probable that some of the lower division may also be present. Sandstones similar to those in the middle section of the Black Pine Range are most widely distributed, especially in the area south of Sublett Creek and east of Lake Fork. Here, too, the sandstone forms smooth slopes only infrequently broken by ledges. Some subordinate beds of limestone occur in it and some of it is cherty, but these beds are all too easily confused with the cherty Permian series which lies above. In several places fossiliferous beds of the lower division appear in lower slopes of stream courses which cross anticlinal structures. In Sec. 36, T. 12 S., R. 29 E., a fauna with *Composita subtilitia* and *Euomphalus ? sp.* was collected and in Sec. 35, T. 11 S., R. 28 E., a fauna with *Composita subtilitia*. Mr. Girty states that these lots may belong together and are more likely Wells than either Brazer or Phosphoria. Although these two localities are more than eight miles apart, they apparently come from the same stratigraphic horizon. The upper white siliceous limestone is also widely distributed. It tends to form knobs along the hillsides or on the ridges and is useful as a marker for the overlying Permian series.

The Wells formation is also well represented in the Albion Range in the low ridge or window between the two main sections of the range east of Oakley. The upper part of the Brazer may be present in the exposures nearest Elba, for there the strata are composed of light gray, poorly fossiliferous limestones which tend to form ledges. Their dip soon carries them from sight below the younger strata and above them appear in turn cherty and sandy limestones much like those in the northern end of the Black Pine Range. These cherty and sandy limestones carry a limited fauna, although crinoid stems are abundant in places. Fossils collected in Sec. 27, T. 13 S., R. 24 E., and identified by Mr. Girty include *Productus cora ?* and *Clriothyridina ? sp.*, which he states may be either Brazer or Wells but more probably the latter. These alternating sandstone and limestone beds carry to the top of the divide from the east and may carry over to the north end of the exposure on the lower western slope of the Albion Range. Sandstones of the middle division occupy most of the west slope of the range. These sandstones are identical to those in the other ranges and like them carry subordinate calcareous members. Again limestone occurs in the upper part, but whether it actually represents the top member or the lower brought into position by faulting was not definitely determined. It is probable that some of the limestone toward the north end of the exposure might be the upper member.

PERMIAN SERIES

Permian strata in this area constitute a single formation, the Phosphoria. This formation is of great economic and scientific interest because in the areas studied by Mansfield in southeastern Idaho it contains valuable and extensive deposits of high-grade phosphate rock, which is the chief mineral resource in that region.

PHOSPHORIA FORMATION

The Phosphoria formation is widely distributed in the Sublett Range, where it lies above the upper limestone member of the Wells formation and is found as bands along the flanks of the larger folds or in the complex crumplings in the folds or along the borders of the faulted areas. Its presence is definitely established on both sides of Sublett

Creek, for several miles on each side of Lake Fork reservoir, and extends into the detached hills of older rock northwest of the main mountainous area. Most of its duplication and widespread occurrence within the area is due to faulting. It is not surely recognized in either the Black Pine or Albion Ranges, but it may occur in both, especially as much chert characteristic of the upper part of the formation occurs on the lower western slope of the middle section of the Black Pine Range.

Although the Phosphoria formation is not of great thickness (never more than 700 feet), it maintains a nearly uniform character over wide areas. Where studied in the region to the east the formation contains two distinct lithologic units, upper and lower. The upper unit, which consists of massive limestones and cherts, is called the Rex chert member, while the lower unit comprises the phosphatic shales, with which are included beds of limestone and of fetid limestone.

Both units appear in the Sublett region, but little was learned of the lower unit. In places the upper limestone of the Wells formation is emphasized by the relative weakness of the beds immediately above, which have retreated, causing knobs and ledges of the underlying rock to form. This relation suggests that soft phosphatic shales may lie above, and that the lower division of the Phosphoria may differ little from that in southeast Idaho. Weak shaly beds were noted in the field, but little significance was attached to them at the time, for it was not until after fossil determinations were made that the Phosphoria was definitely established or recognized. Fossils were collected from a weak shaly member not far above the top of the Wells on the divide between South Heglar Canyon and Shirley Creek in Sec. 8, T. 12 S., R. 29 E. Fossils were abundant in this member, but consisted of a single species, *Pugnoides weeksi*, which Mr. Girty states rather definitely determines the formation to be Phosphoria and probably the shaly part in which beds of phosphate occur. Until further investigation is made in the area, nothing can be said as to the character of the lower part of the Phosphoria. Its thickness is probably not less than 200 feet.

The Rex chert member above is the conspicuous part of the Phosphoria formation, and because of its superior hardness stands out in strong cliffs and ledges. Such ledges are strikingly shown near the reservoir at the junction of Lake Fork and Sublett Creek and on the west slope of one of the outliers northwest of the main range, northwest of Heglar post office. The Rex Chert member is composed of massive chert in the lower part or lower 75 feet, above which are massive beds of gray limestone, some of them crowded with fossils, especially Productus, and also beds of flinty or cherty shale. Some of the upper limestone also contains much chert, which makes it indistinguishable lithologically from the lower part of the Wells. Detritus above the upper division of the Phosphoria is especially strewn with chert fragments. The thickness of the upper unit is probably not less than 400 feet. Cherty facies of the Rex chert member is generally non-fossiliferous, but locally contains sponge spictules and casts of crinoid stems. Collections made of fossils from other members of the Phosphoria formation and submitted to Mr. Girty for identification suggest that the formation is very widespread from Cold Spring Canyon along the border of Raft River Valley northward to the Snake River Plains. It will require detailed study of the entire range to evaluate its true limits and to separate it from the Wells formation. Fossils collected in Sec. 3, T. 13 S., R. 29 E., at the edge of Sublett Creek Valley, contain the following:

Stenopora aff. carbonaria	Pustula sp.
Rhombopora sp.	Productus phosphaticus ?
Schizophoria sp.	Composita subtilitia
Derbya ? sp.	Cleiothyridina ? sp.
Pustula subhorrida	Bellerophon sp.

Mr. Girty states that this collection is also probably Phosphoria, but that it represents the limestone member distinguished in many occurrences, though not here, by containing *Spirifernia pulchra* fauna. A collection from Sec. 10, T. 13 S., R. 29 E., on the other side of the valley and on the opposite side of a fault is also recognized as Phosphoria, but without suggestion as to any definite part. This fauna contains the following:

Leioclema ? sp.	Streblopteria montpelierensis
Pustula subhorrida	Plagioglypta ? sp.

Another collection from Sec. 36, T. 12 S., R. 29 E., is likewise recognized as Phosphoria, but again without suggestion as to any definite part, and contains *Stenopora aff. carbonaria* and *Pustula subhorrida.*

There is strong likelihood that the Phosphoria also outcrops in the middle section of the Black Pine Range, particularly along the lower western slope or the lower foothills tributary to the main high central part. Considerable chert was noted in surface debris, but such material could as well be derived from the lower division of the Wells.

TERTIARY SYSTEM

GENERAL FEATURES

In Cassia County the Tertiary is represented by both sedimentary and volcanic rocks and as these are commonly interbedded, it is difficult to describe the sedimentary alone without mention of the intercalated and overlying lavas. At one time the Tertiary rocks probably mantled the entire region, but even though erosion has subsequently removed considerable of the strata from the later elevated parts of the district, and alluvium has covered much of it in the depressed areas, it remains today one of the most widespread of the geologic formations. The same series of rocks continues eastward into the region where it has been described as the Salt Lake formation, and also to the west where it has been described as typical Payette formation.

Tertiary rocks may be divided into two parts—a lower part consisting wholly of sedimentary strata, and an upper part consisting of alternating beds of lava and sediments. There is no apparent break between the two parts and the sedimentary strata of the upper part are similar to those of the underlying part. Exposed thickness of the Tertiary strata is estimated at 2,500 feet, the upper 800 feet of which constitute the upper division. As a whole the Tertiary rocks are mainly volcanic as much of the sedimentary material consists of bedded ash and tuff.

DISTRIBUTION AND CHARACTER

Best exposures of the Tertiary strata are in western Cassia County at the margin of the district where Bowen[1] has described the occurrence in the lower division of clay, shale, volcanic ash, sandstone, conglomerate, and some lignite, all well stratified and for the most part regularly bedded. Clay and shale predominate in the lower part, volcanic ash and sandstone in the upper part. The ash beds consist of white or gray ash of pumiceous character. The sedimentary members of the upper division are similar to those of the upper part of the lower division, but are intercalated with flows of acidic lava. A composite section of the Tertiary strata measured along Trapper Creek west of Oakley is reproduced below from Piper's[2] report on the Goose Creek Basin:

GENERALIZED COMPOSITE SECTION OF LATE MIOCENE (?) ROCKS MEASURED
ALONG TRAPPER CREEK WEST OF OAKLEY

Member	Thickness (feet)
1. Rhyolite[3] cap rock	101
2. Volcanic ash or "lake beds," gray to white, bedded, friable	12
3. Rhyolite	115
4. Volcanic ash, gray to white, stratified, locally crossbedded, poorly consolidated	132
5. Rhyolite	28

[1] Bowen, C. F., Coal and lignite in Boise and Cassia counties, Idaho: U.S. Geol. Survey Bull. 531-H, 1913, pp. 252-262.
[2] Idaho Bureau of Mines and Geology Bull. 6, op. cit., p. 28.
[3] Microscopic studies by Mr. E. S. Larsen show that the rhyolites are actually quartz latites, Bull. 6, op. cit., pp. 27, 30.

6. Volcanic ash showing local crossbedding. One bed of buff sandy clay grading downward into typical ash...................................... 120
7. Rhyolite... 68
8. Volcanic ash beds, outcrops masked by talus..................... 410
9. Rhyolite... 56
10. Volcanic ash, gray, coarse... 29
11. Lignite and alternating fine white clays, some intercalated ash. Five lignite beds 0.2 to 1.0 foot thick................................ 30
12. Two strata of fine white clay grading downward into typical volcanic ash.. 255
13. Lignite, enclosed by fine white clay.............................. 2
14. Volcanic ash, clay, sandy clay, and ashy sandstone............... 232
15. Dense white to blue clay, several intercalated streaks of carbonaceous shale... 7
16. Conglomerate, loosely consolidated, pebbles of limestone, chert, and quartzite... 4
17. Fine sandy clay, clayey ash, fine blue shale, and typical volcanic ash interbedded.. 12
18. Conglomerate, sandy at top. Pebbles of limestone, chert and quartzite 12
19. Volcanic ash, with some intercalated argillaceous beds, to bottom of exposed section....................................... 75+

From this section it is readily seen that the series may be divided into two parts representing two periods differing in the intensity of the volcanic activity. The upper division includes members 1 to 9 of the composite section and comprises the lava flows and intercalated beds of poorly consolidated volcanic ash. The lower division includes the remainder of the section, but no marked difference in character separates the two.

Some of the sandstone and conglomerate in the upper division of the series outcrops in the canyons at the west base of South Mountain, but elsewhere in the district the sedimentary strata are largely concealed by the upper lavas. The strata at the west base of South Mountain are continuous with those of the Goose Creek region.

The Malta Range is carved wholly in the Tertiary series, but most of the sedimentary strata are concealed beneath the capping lava flows. Here and there near the west base of the tilted block range the gullies have cut through the lavas and have exposed the underlying stratified rocks. Most of the disclosed strata consist of tuff or ash of pumiceous character, resembling powdered and semi-consolidated pumice and containing variable amounts of plagioclase and augite fragments, and magnetite grains. The tuff or ash is regularly bedded, as shown in Plate XVI, A, and the difference in the individual layers is mainly in the coarseness of the various constituents and also in their relative proportions. A much thicker sequence is shown along the east front of the range, but unfortunately exposures are largely obscured by landslide blocks of the cap lava and such strata as are exposed at the base of these blocks are wholly inadequate for study. Locally, near the south end of the range, about 400 feet of the strata are exposed, wholly of pumiceous material. Much ash, but including some clays and shales, also form low hills in the wide basin south of Almo, and appear also at the base of the Raft River Range. Some of the shale is well compacted and has greenish and bluish colors.

Some remnants of the tuff and shale also lie at the north end of the Albion Range in places beneath the lava. This sedimentary rock belongs to the upper division of the series and its position suggests that it was originally deposited on a nearly level surface or the old erosion surface coextensive with that now exposed on the summit of the range near Mount Harrison and Cache Peak.

Most of the upper part of the series with intercalated lavas and lava cappings has been eroded from the area east of the Raft River Valley, leaving only remnants of some of the lowest flows and the underlying sedimentary strata. Much white or gray pumiceous ash and also dark gray basaltic tuff, the latter composed mainly of well-compacted brownish glass and plagioclase and augite fragments, remain in parts of the Sublett Range and in the wide valley between it and the Black Pine Range. Much of this is beneath a thin flow of dark glassy lava. Along lower Sublett Creek these strata contain an intercalated flow of basalt. Members still lower in the series appear at the north end of the Black Pine Range. These are more typically sedimentary and consist of conglomerates and marls or fresh-water limestones. The conglomerates are relatively soft and loose textured, and are composed of angular to subangular and even rounded pebbles. These are composed of local materials and the pebbles are usually sandstone, limestone, and chert from the Wells formation in a calcareous or sandy matrix. These beds also appear in the southern slopes of the Sublett Range. Beds of marl or fresh-water limestone, measuring about 100 feet thick, outcrop at the north tip of the Black Pine Range under a partial capping of white volcanic ash. Fresh water gastropods are numerous in the marl. A thin bed of conglomerate underlies this member.

Exposures suited for measurement were not observed east of the Raft River Valley, but at least 500 feet of beds, including lavas, were seen at the surface. The total thickness is probably much greater, for the strata above have accumulated on and largely buried a topography of considerable and varied relief. East of Raft River Valley the strata are much like those of the Salt Lake formation as exposed in the Fort

Hall Indian Reservation[1] a few miles to the east. One is indistinguishable from the other in lithologic character and manner of deposition, and both have the same association with igneous rocks.

CORRELATION AND AGE

Definite correlation of this series is not wholly possible, as the only fossil shells found in it consist of fresh-water gastropods similar to those found farther east in the Salt Lake formation. Unfortunately, these shells have no diagnostic properties, for they range throughout the Tertiary. Stems and broken fragments of plants have been found in several places west of Oakley associated with lignite, but no specimens were found which could be identified.

CORRELATION WITH THE PAYETTE FORMATION

The strata have striking lithologic resemblance to the Payette formation in southwest Idaho, in which abundant Upper Miocene fossil plants have been found. Bowen,[2] and later Piper,[3] attempted correlation of the strata in the Goose Creek area with the Payette beds, but were not wholly successful because of early confusion in the designation of the Payette horizon. Both concluded that the Goose Creek strata must be older than Payette. Subsequent work has corrected the early error. To show its bearing on the general problem, the earlier work will be briefly reviewed.

As first described by Lindgren,[4] the Payette consists of a thick series of lake beds composed of shales, clays, sands, and tuffs of white or light color above a series of lavas dominantly basalt but with rhyolite at the top. Fossil plants in the sedimentary series were assigned to the Miocene and to the Pliocene, indicating two series of sedimentary strata, nearly indistinguishable lithologically, to the younger of which he gave the name Idaho formation. In 1904, while working in the Owyhee Mountains of Owyhee County, he[5] assigned the Payette to the Eocene on a revision made by Knowlton on the Payette flora.

Bowen[6] cited the resemblance of the Goose Creek beds to the Payette formation of the Snake River Valley region in lithologic character, degree of consolidation, and manner of deposition; and also the associa-

---

[1] Mansfield, G. R., Geography, geology, and mineral resources of the Fort Hall Indian Reservation, Idaho: U.S. Geol. Survey Bull. 713, 1920, pp. 54-55, 70.

[2] Bull. 531-H, op. cit., p. 14.

[3] Bull. 6, op. cit., p. 35.

[4] Lindgren, Waldemar, Description of the Boise quadrangle: U.S. Geol. Survey Geol. Atlas, Folio 45, 1898.

Mining Districts of the Idaho Basin and the Boise Ridge, Idaho: U.S. Geol. Survey Eighteenth Annual Rept., pt. 3, 1898, pp. 625-736.

[5] Lindgren, Waldemar, Description of the Silver City quadrangle: U.S. Geol. Survey Geol. Atlas, Folio 104, 1904.

[6] Bull. 531-H, op. cit., p. 14.

tion with rhyolites similar in all respects to those in the Snake River Valley; all of which have been regarded by Lindgren and others as of early Tertiary age. He stated:

"It cannot be said, however, that these strata are the equivalent of the Payette formation, because (1) in the Snake River Valley the rhyolite is older than the Payette beds, whereas in the Goose Creek district the rhyolite is interstratified with the upper division of the sedimentary rocks, which are chiefly older than the rhyolite; (2) the Payette formation in the Snake River Valley is associated with basalt. In the Goose Creek plateau no basalt is present and the rhyolite is older than the basalt of the Snake River Plains adjacent on the north. The Goose Creek district was occupied by a lake prior to the rhyolite extrusion. If the rhyolite of the Goose Creek district and that of the Snake River Valley are of the same age, the Snake River Valley, to which Goose Creek is tributary, must also have been occupied by a lake prior to the advent of the rhyolite and therefore before the deposition of the Payette formation. This would mean either that the old lake in which the Payette beds were laid down was in existence earlier in the Eocene epoch than the deposition of these beds or that an earlier lake whose history is not yet known occupied the valley. No evidence of the existence of such a lake has yet been observed in the Snake River Valley.

"If, on the other hand, the rhyolite of the Goose Creek district is younger than that of the Snake River Valley, the sedimentary strata of the two localities may be of the same age and all belong to the Payette formation."

Buwalda[1] in a more recent work in southwestern Idaho established the existence of two series of tuff beds on the basis of vertebrate fossils found in the area covered by Lindgren in the Silver City geologic folio. From the field identification of fossil teeth, the beds in the vicinity of Rockville, near the western boundary of the Silver City area, are termed middle to late Miocene and are *older* than or contemporaneous with the rhyolite flows of the region. He then inferred that the lithologically similar beds which border Snake River farther east and called the Payette formation by Lindgren are definitely younger and of a late Miocene age or more recent.

With the new data as to the age of the Payette formation Piper[2] stated his belief that the Goose Creek series was contemporaneous with the rhyolites of the Snake River Plains and the Silver City region and

[1] Buwalda, J. P., A preliminary reconnaissance of the gas and oil possibilities of southwestern and south central Idaho: Idaho Bureau of Mines and Geology Pamphlet 5, 1923, p. 2.
[2] Bull. 6, op. cit., p. 35.

older than the Payette formation described by Lindgren or equivalent to the Upper Miocene strata mentioned by Buwalda. His assignment differs from Bowen's only in that the series belongs to the Miocene instead of the Eocene.

More recently Kirkham,[1] who has restudied the Tertiary strata in southwestern Idaho, redefined the Payette and Idaho formations and assigned the Payette formation to the series of strata beneath the rhyolite and the Idaho formation to the strata above. The Payette sediments beneath the rhyolite contain Upper Miocene flora and are contained between the flows of Columbia River basalt. The Idaho lake beds above the rhyolite yield a Pliocene flora.

In light of these most recent findings there is no difficulty in correlating the sedimentary series with its intercalated and capping flows of lava in Cassia County with the Payette formation farther east, since both lie below the rhyolite horizon. The rhyolite or acid lava cap is the same in the two regions, for the writer as well as Kirkham[2] has traced the capping flows westward from Cassia County into the region described by Lindgren in the Silver City folio and has found them continuous. The most notable change from southwestern Idaho to the south and southeast part of the State is the great decrease in the number of Columbia River basalt flows in the Payette and the corresponding increase in the number of acid lava flows in the same stratigraphic horizon or above.

CORRELATION WITH THE SALT LAKE FORMATION

Statement has already been made of the similarity of the Tertiary series in Cassia County to the Salt Lake formation in the Fort Hall Indian Reservation. The lower Salt Lake strata are mainly sedimentary composed of conglomerates, marls, limestones, and clays, overlain by white or gray pumiceous ash or tuff and containing and capped by flows of acidic lava. This general sequence is the same as in Cassia County and the strata in the two districts are indistinguishable.

Mansfield[3] has assigned the Salt Lake formation tentatively to the Pliocene on the relation of the acid lavas in its upper part to the rhyolites in Yellowstone National Park. Such fossils as have been found in the series have not been of determinative value or are of long-ranging types. In Yellowstone Park the "Canyon" conglomerate, which, according to W. C. Alden, may prove to be younger than the main body of the rhyolite, has yielded fossil remains that were identified by

---

[1] Kirkham, V. R. D., Revision of the Payette and Idaho formations: Journal of Geology, Vol. XXXIX, 1931, pp. 193-239.
[2] Kirkham, V. R. D., Oral communication.
[3] Prof. Paper 152, op. cit., pp. 111-112.

Prof. O. C. Marsh as belonging to the skeleton of a fossil horse of Pliocene time.[1] These rhyolites, according to Mansfield, extend westward from Yellowstone Park along the northern parts of the ranges into the Fort Hall Indian Reservation, not absolutely continuous, but the occurrences are so numerous and the characters of the rocks so similar that there seems little reasonable doubt that the rhyolites in that region and the Yellowstone National Park are of essentially the same geologic age. He mentions that the rhyolite and the Salt Lake formation are here interbedded and associated with beds of volcanic ash.

Similar views are shared by Kirkham,[2] who shows that the Salt Lake formation with intercalated and overlying flows of lava extends almost continuously about the margin of the upper end of the Snake River Plains and that the upper flows are continuous with those in Yellowstone National Park. These relations have been observed by the writer, who has also traced the flows to the Park from the region near the Fort Hall Indian Reservation.

Outcrops between the Fort Hall Indian Reservation and Cassia County are not absolutely continuous, but occur at sufficiently close intervals in the region to make it certain that they belong together and that their separation is due only to erosional stripping. Because the stratigraphic relations, lithology, and other factors are so similar between the series in Cassia County and the region described by Mansfield and Kirkham and because the volcano-sedimentary series is beneath similar essentially continuous sheets of acidic lava, the suggestion is very strong that the two are of the same age and remnants of a former continuous blanket deposit. Since the present report was written,[3] Kirkham has further suggested that although no definite flora or fauna from the Salt Lake formation has as yet been determined, it may, from its relations to the rhyolite, eventually prove to be contemporaneous with the Payette formation. The Salt Lake formation would thus be equivalent to the Payette formation and belong in the Middle and Upper Miocene instead of in the Pliocene epoch. Blackwelder,[4] who has studied the Salt Lake series in the region of the Wasatch Range in Utah, also considers the formation to be not younger than Miocene.

QUATERNARY SYSTEM

The rocks of the Quaternary consist for the most part of unconsolidated sand, gravel, and mud that occupy the floors of the larger

1 U.S. Geol. Survey Geol. Atlas, Yellowstone National Park folio (No. 30), 1896.
2 Kirkham, V. R. D., Pamphlet 19, op. cit., p. 34.
3 Kirkham, V. R. D., Snake River Downwarp, Jour. of Geology, Vol. 39, 1931, p. 472.
4 Blackwelder, Eliot, Wasatch Mountains revisited: Bull. Geol. Soc. Amer. abstract, Vol. 36, 1925 pp. 132-33.

valleys and the broad aggraded desert basins between ranges; and of coarser or finer debris, less well stratified, which has accumulated here and there along the lower slopes of the mountains. Most of it is the result of fluviatile action where the products of rock disintegration and decay have been washed from the valleys and slopes of the mountains into the structural depressions and basins, its deposition aided in large part by damming of these basins by flows of Snake River basalt. Some of it, however, is the product of glacial activity, but these deposits are not nearly so extensive as the others, and have for the most part been masked by the younger debris carried down from the same localities. In some places aoelian deposits are extensive.

The Quaternary deposits are all unconformable upon older rocks and range in age from Pleistocene to Recent. It was not found practical to separate them in mapping.

PLEISTOCENE SERIES

The Pleistocene series includes most of the Quaternary deposits, although it is not always easy to draw a line between them and the deposits of more recent age. Most of the waste now in the broad basins is unquestionably Pleistocene in age and its deposition may even have started in the Pliocene. The extensive and thick alluvial-fan deposits at the base or on the lower slopes of the mountains were mainly completed during the Pleistocene and have received only minor contribution since. Terrace deposits in connection with old Lake Bonneville are also Pleistocene. Deposits of two glacial stages are present, but morainal forms of the earlier have been much subdued and mainly concealed, whereas those of the younger, although fresh and unmodified, are only locally abundant. For convenience the Pleistocene deposits will be described as hill wash and older alluvium, glacial deposits, and aeolian deposits.

HILL WASH AND OLDER ALLUVIUM

The lower slopes of most of the hills and ranges are covered with rock fragments and soil which effectively conceal the underlying formations. In the main this material is poorly sorted or without definite arrangement, but in places it merges with fluviatile deposits and becomes a part of the vast alluvial plain at the base of all the ranges. The upper limit of the deposits, as mapped, is the general line that marks the outcrop of the older, underlying formations, or the line where the debris of these rocks is sufficiently characteristic and abundant to indicate clearly the strong probability of their occurrence beneath. The line is much generalized, especially where it lies between the Tertiary sediments and hill wash. It was especially difficult to mark the boundary between the alluvium and the older rocks about the margin of the Sublett Range and in some places a line taken a mile or two on either side would have served as well.

Near the borders of the mountains the deposits have been slightly trenched and the surface has been made somewhat irregular, but away from the base of the ranges the alluvial plain has a smooth surface, in places floored with gravel, but usually concealed by aeolian soil. The materials of the hill wash and older alluvium are mostly local or derived from nearby sources and are composed of pebbles and sands of quartzite, sandstone, limestone, and chert. On the south slope of the Black Pine Range the hill wash and alluvial-fan material are relatively thin. The thickness in the center of the basins could not be obtained, but probably ranges from hundreds of feet in the smaller basins to a thousand feet or more in the center of Raft River Valley.

Terrace deposits are extensive only on the southeast and east slopes of the Black Pine Range, where they form a series of sloping beaches, each corresponding to a stationary period in the development or history of old Lake Bonneville. From the position of these terraces, in part upon alluvial fans or the piedmont alluvial plain, it is clear that the hill wash and piedmont deposits were largely completed in the pre-Bonneville epoch and so are of Pleistocene age.

OLDER GLACIAL DEPOSITS

Moraines and outwash of an early epoch related to the carving of the broad U-shaped valleys in the Albion and Raft River ranges and South Mountain are poorly preserved. Such deposits have in large part been smoothed over and largely concealed by younger deposits. Glaciers from the basin between South Mountain and the Albion Range extended to the edge of the plain a short distance south of Oakley and left there an outwash plain of fine materials along with occasional quartzite erratics. At the same time ice also crossed the Albion Range and deposited moraine in the Almo Basin several miles from the edge of the mountain. In a few places irregular topography still obtains, although greatly reduced from its former dimensions because of filling from the higher slopes. Glaciated quartzite boulders are common, but granite has apparently long since weathered away. The outwash merges and is mainly indistinguishable from the other alluvial deposits in the Basin. Much outwash was also deposited on the south side of the Almo Basin derived from glaciers in the Raft River Range.

Glacial deposits may be recognized in other parts of the Albion Range, particularly on the east side. Moraine from both the Malta and Albion ranges end in steep banks within a few dozen feet of one another south of Elba. Erratics rest high on the alluvial slopes bordering Conner Creek. Large angular quartzite boulders are common in the Albion Basin near the foot of the mountain southwest of Albion. Only glaciers could transport such boulders and carve such wide U-shaped valleys in the resistant quartzites of the range. Most of the basin filling was probably accomplished before or during the glacial epoch. In places erratics occur on the west side of the range, but glaciation was not nearly so extensive on that side.

Extremely high and steep alluvial slopes occur on the northeast side of the Black Pine Range, now deeply trenched by valleys issuing from the range. These abnormally steep and partially dissected slopes might represent moraine and outwash from early glaciers, whose present surface is stranded about 500 feet above the present valley floors at the edge of the mountain. A lower terrace in the trenched alluvium might record the outwash of younger glaciers; this terrace now 200 feet above the present valley floor. A few miles to the east these high terraces descend to the level of the Lake Bonneville terraces. Unfortunately, the gravel in these high terraces stands in steep smooth slopes and it was not found possible to examine the deposits in detail or learn much of their character.

YOUNGER GLACIAL DEPOSITS

The only glacial deposits of younger age were definitely recognized in the Albion Range high above the alluvial basins and usually not below 7,500 feet A.T. These deposits are characterized by extreme freshness in form and materials and occur below the glacial cirques near the top of Mount Harrison and below Independence Lakes on the north side of Cache Peak. Lake Cleveland is held in large part behind a morainal dam. Deposits continue only a short distance below the lake. Moraine is more extensively developed below Independence Lakes and not only confines the five little lakes in the basin at the head of Green Creek, but actually descends to the very foot of the mountain on a gradient so steep that from a distance the morainic forms appear to overlap one upon the other. Subsequent drainage has done very little to modify the form of the drift and outwash.

AEOLIAN DEPOSITS

Windblown deposits are conspicuous only in the northeast part of the district in the broad basin between the outlying hills and the main

Sublett Range at Heglar, at the north end of the range near the Snake River, and to a lesser extent on the slope east of Sublett post office. A smaller area also lies on the east slope of the Albion Range north of the town of Albion. Areas of aeolian deposits may be distinguished as the only places where dry farming has been successful, and, unless the slopes are too steep, are marked by summer fallow or wheat fields. They are found mainly on the leeward side of ridges or hills where check in wind velocity has permitted settling and accumulation of dust.

The deposits are typically loess, in color and texture and in ability to stand in vertical banks. Lack of horizontal lamination and a distinct vertical parting also favors their wind-born origin. North of Heglar the loess has been submaturely dissected and the surface is closely similar to the rolling Palouse Hills in the northern part of the State. The deposits show great range in thickness from a mere film where the wind sweeps across the ridges to 50 or 100 feet or more on the east slopes and to even greater thicknesses on the floors of sheltered basins. Along the Snake River, a mile or two to the north of the area, loess banks a hundred feet or more high may be seen from the highway. The loess deposits near Albion are not so thick and do not entirely conceal the older formations below.

The source of the loess is in part from the fine silts blown from near the centers of Raft River and Goose Creek valleys and carried to the northeast by the prevailing winds, but most of it is probably of glacial origin and has its source in the outwash of the early glaciers. Both areas of loess accumulation are in the direct path of the prevailing winds from the main centers of glaciation in the south end of the Albion Range and the outwash plain near Oakley.

RECENT SERIES

ALLUVIUM

In the broader valley bottoms and along the courses of so many of the larger streams, such as Raft River, Marsh Creek, Sublett Creek, and Cassia Creek, are narrow bands or broader areas of commonly well sorted and fine or locally gravelly flood plain deposits. In some places these merge gradually with the older alluvium plain, but more generally they lie between low bluffs.

Landslides also are abundant, not only in the Malta Range where the volcano-sedimentary series has especially prepared an unwieldy base for such movement, but also in the Sublett Range, where the oversteepened valleys of the Blackfoot cycle have favored mass-movement of debris into the valley bottoms. In places in the Sublett Range valleys have been partially filled by these slides.

IGNEOUS GEOLOGY

GENERAL OUTLINE

The igneous geology in eastern Cassia County is represented by intrusions of granitic rock and by extrusions of acid lavas and basalt. These have occurred in different epochs in the Tertiary period, beginning first with the intrusion of plutonic bodies either late in the Cretaceous or early in the Eocene. Next occurred the extrusions of latites and perhaps other acid lavas, locally basalt, during the late Payette epoch, probably late in the Miocene or early in the Pliocene. Last was the extrusion of the Snake River basalt rather late in the Quaternary. Each of these has followed certain periods of structural unrest and each is in no way related to preceding or following periods of volcanism.

These rocks are less widely distributed in eastern Cassia County than the sedimentary rocks and less widespread than in adjoining areas on the west and north. Very likely Cassia County was at one time entirely blanketed by lavas of the Tertiary epoch, but due to later mountain-building movements such rocks have been largely eroded from the higher lands and concealed beneath alluvium in the depressed areas. Nevertheless, large areas of each kind remain within the district. Flows of Snake River basalt were added well after the development of the present topographic features of the district.

INTRUSIVE ROCKS

DISTRIBUTION AND EXTENT

Intrusive rocks occur as a large stock or small batholith, to which the name Cassia batholith has been given, as smaller stocks or cupola-like masses which comprise higher parts in the roof of the major batholith, and by smaller dikes of associated pegmatite. These cover more than 60 square miles, mainly in the southern part of the Albion Range and in South Mountain. The Cassia batholith has its southern end about two miles north of the Utah line and forms the core for most of the south part of the Albion Range. It extends northward over Cache Peak and far down its western and northeastern flanks. Outliers extend as far north as Conner Creek and Cottonwood Creek in the northern section of the Range and appear frequently along the west slope of the southern section. The two most northerly stocks are in valley walls and are undoubtedly continuous beneath the quartzite capping which separates the two valleys. Those on the west flank of the range represent higher cupola-like bodies rising above the larger Cassia batholith and are continuous with each other and with the greater body beneath a thin shell of the pre-Cambrian quartzite series. The granitic bodies in South Mountain appear mainly in the steep eastward facing scarp beneath a thin capping of the pre-Cambrian strata. Outliers here, too, appear along the more gentle westward slope. This granitic body is a part of the main Cassia batholith in the Albion Range and has been brought to its present position overlooking the intervening basin by comparatively recent faulting, which has separated the South Mountain block from the Albion Range. Had erosion continued everywhere to an additional depth of 500 feet the granitic rock now exposed at the surface would probably have been more than doubled.

Erosion has carved more deeply into the eastern side of the Albion Range, especially south of Cache Peak, and has there exposed much more of the batholith. When once the resistant quartzite shell was pierced erosion in the granite was comparatively easy and rapid and the elongated basin-like area that contains the Cassia City of Rocks was carved. From this basin an excellent vertical section of the batholith may be had. Its walls plunge somewhat more steeply than the slopes of the range and the dips of the overlying sedimentary strata, and the body thus grows larger with depth. More than 5,000 feet of granitic rock is exposed from the top of Cache Peak to the bottom of the basin.

STRUCTURAL FEATURES

The Cassia batholith has the characteristic features usually assigned to batholiths; that is, it forms a plutonic mass exceeding 40 square miles in surface area. It is dome-shaped and has steeply plunging walls, without apparent bottom. In places its contact with the invaded strata is sharp, but more generally there is a transition zone between invading and invaded rock. This zone is usually not more than 100 feet wide and seldom more than 50 feet. Through this interval the granitic rock appears to merge gradually with the quartzitic strata and it is usually impossible to fix a precise boundary between the two formations. Much of the outer part of the batholith and the stocks has a gneissic banding and this structure is carried into the sedimentary strata, adding to the difficulty of separation. Such transitory relation suggests a marked infiltration of magmatic material from the igneous body into the invaded strata during the intrusion. In most places the transition zone is slightly darker in color than the rock on either side, due mainly to enrichment in biotite.

Jointing is a pronounced structure in some parts of the batholith and its outliers, and locally plays an important role in the origin of the curious forms in the Cassia City of Rocks. Much of the batholith in the City of Rocks area is cut by widely spaced joints, often outlining blocks as much as 30 to 40 feet across. In places, however, the joints are more closely spaced and locally produce a sheeted structure in the granitic rock. Usually the jointing is not noticeable except where weathering has accentuated the joint fractures. One set of joints trends to the northwest, another to the northeast, and a third set is nearly horizontal.

PETROGRAPHY

GRANITIC ROCKS

The batholith and its outliers are composed of rock ranging from granite to granodiorite in composition, with granite usually as the outer border of the bodies and the granodiorite as the great inner core. This order of gradation is the reverse of that usually found in plutonic bodies, for normally, granodiorite comprises the marginal zone and the more acid rock the inner part. The two main facies may in part be distinguished by texture. The granodiorite has an even grain, whereas the granite is porphyritic and usually gneissic. In this way the batholith shows a progressive change from a medium to coarse even grained rock in the central part to porphyritic rock outward; at first faintly gneissic but becoming increasingly more gneissic toward the margin, passing locally to augen gneiss. Smaller bodies differ from the larger only in having the characteristics of the outermost part and have rarely the composition or texture of granodiorite.

Most typical of the granodiorite facies is the core of the batholith in which the Cassia City of Rocks is carved. Near the south end of the Rock City the granodiorite is light gray to nearly white and is of medium to medium-coarse grain. Smoky grains of quartz are numerous and are in a chalky white feldspathic mass, together with a liberal sprinkling of partially altered biotite flakes. In common with most of the rock in the area, it is friable as a result of granular disintegration and is readily reduced to sand. Its texture is typically hypautomorphic granular in thin section and its dominant constituents are oligoclase (53%), quartz (25%), microcline (12%), and biotite (7%), and its accessories include magnetite, zircon, and apatite. The oligoclase shows considerable zoning and has interiors as basic as acid andesine. Crystals are always twinned polysynthetically and this twinning is occasionally combined with that of the carlsbad and pericline laws. The plagioclase is much altered to large flakes of secondary white mica and has occasional seams or veins of a clay mineral. The latter is a product of surface weathering, but the development of muscovite is the result of a deep-seated process. The microcline forms large enveloping grains and shows from its cementing relations that it was the last mineral to complete its crystallization. Its twinning is usually indistinct. In sharp contrast with the plagioclase, the microcline shows only slight alteration to kaolin and sericite. Biotite and quartz show no unusual characters, the biotite occurring in scattered leaves or flakes partially altered to chlorite, and the quartz in rather large anhedral grains. Zircon are unusually plentiful. In addition to the secondary product already given, are epidote and zoisite.

Near the head of Circle Creek the rock shows some notable differences, mainly in a greatly increased muscovite content and locally in the development of tiny reddish garnets. This rock is also of medium grain, has a light gray color and is composed entirely of light minerals. It has the usual hypautomorphic granular texture and is composed essentially of quartz (44%), acid andesine (30%), microcline (12%), muscovite (8%), and garnet (3%), and with magnetite, apatite, and zircon as additional accessories. Quartz is exceptionally abundant, but the batholith as a whole is comparatively rich in silica. There has been no change in the mineral relationships, but occurrence of so much muscovite and garnet suggests that pneumatolytic action has been prominently connected with the intrusion. Muscovite is prominent throughout much of the body, accompanied or unaccompanied by biotite, and for the most part the inner core might be classed as a muscovite-biotite granodiorite.

Toward the margin the rock maintains essentially its same coarseness of grain, but scattered phenocrysts appear and these form much larger sizes. Most of the phenocrysts are from one-fourth of an inch to one inch long, but some attain a length of as much as six inches. With this change in texture most of the rock also assumes a faint gneissic banding, which becomes increasingly more pronounced in the outer marginal zone. In all places examined the marginal rock has the composition of granite with the large phenocrysts composed of microcline, but the other minerals are essentially those in the granodiorite. A specimen from the stock on Cottonwood Creek in the north section of the Albion Range is representative of the non-gneissic facies. It shows numerous large microclines in a ground mass of medium grain composed of quartz, oligoclase, biotite, and muscovite. The microcline is the most striking, for it tends to hold an idiomorphic outline, yet its outer borders are minutely irregular where it penetrates and partly wraps around other minerals. Often it holds rounded grains of other minerals, particularly quartz and plagioclase. Otherwise the rock is no different from the granodiorite, except for its greatly increased microcline content. The oligoclase has altered almost entirely to a sericitic mass, but the microcline shows scarcely a sign of alteration. Quartz is abundant and tends to form granular masses. Accessories are zircon, magnetite, apatite, and rutile.

Specimens from the stock on Conner Creek and from the north side of Cache Peak are representative of the gneissic prophyritic granite. These differ from the rock just described in that the sections are streaked with finely granulated zones of quartz and feldspar and with similarly disposed zones or lenses of coarse granular quartz in which biotite plates or flakes are arranged parallel to the banding or schistosity in much the same way as in a recrystallized micaceous quartzite. Except for the granulated zones or cataclastic structure and the more or less pronounced parallel orientation of the minerals, the rocks have the same list of minerals in the same proportions and in the same relations as in the non-gneissic variety. Alteration of the oligoclase to secondary white mica is probably more pronounced, but the microcline remains perfectly clear. The biotite occurs only in the quartz masses. Accessory minerals as well as secondary products are the same as before.

The gneissic shell or transition zone has essentially the same composition as gneissic granite, but its gneissic structure is much more pronounced and cataclastic structures are much more evident. The zone is mostly represented by a coarse grained strongly banded lenticular gneiss composed of alternating quartzose and feldspathic lenses or seams, the quartzose parts rich in biotite. Some layers also alternate with seams of nearly pure pegmatitic material. As in the gneissic granite the quartz occurs mainly as granular aggregates in lenses or seams with parallel orientated biotite, but the microcline and plagioclase lie in any position in the border zones. In general the microcline predominates, but locally it is subordinate to the oligoclase. The microcline invariably holds numerous inclusions or remnants of quartz and oligoclase and also shows less alteration to white mica than does the oligoclase. Large flakes of muscovite are commonly present and also the usual accessories of the granite, such as zircon and magnetite. Garnet also appears as an unfailing constituent, but it is always very subordinate. The rock continues to show increasing amounts of quartz as it nears the little altered quartzite. Large muscovite flakes and minor amounts of feldspar continue in most places for some distance into the quartzite. Locally, instead of such a feldspathic shell, the border is dark colored and composed largely of hornblende, with lesser quartz, epidote, titanite, oligoclase, microcline, and zoisite.

Such changes in the texture of the rock from massive in the center to gneissic at the margins of the bodies may perhaps be explained on the assumption of powerful lateral pressures during the crystallization of the rock, a process designated by the term piezocrystallization. During such action, the plagioclases are commonly said to be broken down to form microcline.

PEGMATITES

Pegmatite dikes and sills are not especially numerous about the main batholith and most of the stocks, but they do show in great abundance in the upper part of the granite and the overlying shell in South Mountain and attain their maximum development in the southern part of the mountain near the Utah line. These apparently constitute the only differentiates of the deep-seated granitic masses, unless the quartz veins in the north should also be so regarded.

The pegmatites range from thin seams to lenticular bodies 50 feet or more across and several hundred feet long. In general they have the texture of coarse-grained granites, but some are composed almost wholly of graphic granite. The minerals of the pegmatites are not especially large, although some of the pegmatites near the Utah line have been prospected for mica, but neither the muscovite nor the biotite crystals exceed two or three inches in diameter.

Specimens from a large pegmatite on the upper slope of South Mountain and typical of all the pegmatites in the district were examined microscopically. The body is composed of a light gray or white coarse-grained rock, in part coarsely granular but in part graphic, made up of milky white feldspar, grayish quartz, and muscovite. Some of the muscovite flakes are two inches or more across, but much of it is in smaller scales and crystals. The feldspar is also less than two inches, but locally at least as in the Cassia City of Ricks, crystals six inches long have been weathered out and occasionally are strewn on the surface. In addition to the minerals listed the microscope reveals minor amounts of garnet, zircon, apatite, titanite, rutile, magnetite, with occasionally columbite and sericite. The feldspar is oligoclase, approaching albite in composition, and is invariably intergrown with orthoclase as antiperthite. Some of the quartz is a late introduction and forms veinlets cutting the other minerals.   One such veinlet contains scattered garnets.

The pegmatites have apparently had a profound influence on the metamorphism of the quartzites, for the transitional shell attains its greatest thickness in areas where pegmatites are most abundant.   In these areas large flakes of muscovite are also developed in the quartzites long distances from any observed granite rock.   The metamorphism is undoubtedly the result of the transference of volatile constituents from the intrusive magma and such volatile matter would be especially concentrated in the areas where pegmatites are now most abundant.

AGE OF THE INTRUSIVES

The Cassia batholith and its outliers may be assigned with considerable assurance to the late Cretaceous or early Tertiary and may be correlated with the intrusives of that age in the Rocky Mountain region rather than with the late Jurassic intrusives of the Coast region.   It is believed that the main epoch of folding and faulting in southeastern Idaho[1] has been a part of the great mountain-building disturbance that came in the interval between the end of the Cretaceous period and the early part of the Tertiary, generally known as the Laramide revolution. At this time the great low-angled overthrusts, such as the Bannock overthrust in southeastern Idaho, and the Albion overthrust in Cassia County, came into being.   The Albion Range is carved mainly in an overthrust block of pre-Cambrian sediments resting upon Carboniferous strata.   Unfortunately the intrusive bodies were not seen in contact with the Carboniferous rocks, but veins related to the intrusive pass from the younger to the older overlying rocks, proving that the mineralization followed the completion of the overthrusting and that the granitic bodies themselves are probably younger than the overthrusts. Such relation favors an early Tertiary date for the intrusion and links it with the early Tertiary intrusions in Utah.

The Cassia batholith is probably an outlier of the well known Idaho batholith.   Although the Idaho batholith has generally been regarded as of late Cretaceous or early Eocene age, Ross[2] considers the evidence rather indefinite and suggests that the batholith is probably as old as late Jurassic or early Cretaceous, or the same age as the Coast Range batholiths.   Should the Cassia batholith actually be an outlier of the Idaho batholith, as the writer strongly believes, the evidence in Cassia County would tend to affirm the earlier assignment of the intrusives to

[1] Mansfield, G. R., Prof. Paper 152, op. cit., pp. 169-170.
[2] Ross, C. P., Mesozoic and Tertiary granitic rocks in Idaho: Jour. Geology, Vol. 36, pp. 673-693, 1929

the late Cretaceous or early Eocene. That the Cassia batholith does not belong to the younger granitic rocks of Miocene age (?), recognized elsewhere in the State,[1] is certain, for the Upper Miocene or Lower Pliocene lavas rest on the eroded surface of the body in the City of Rocks area.

WEATHERING OF GRANITE

Weathering of the granite rock in the district takes on an added interest, for in it lies the explanation of the weird and fantastic forms in the Cassia City of Rocks. The weathering of the granite here is not vastly different from the weathering of similar rock in arid and semiarid regions elsewhere, but the results obtained are far more striking. Most of the weathering may be ascribed to a form of granular disintegration in which the granite is reduced to an arkosic sand of remarkably fresh appearance. Studies of the disintegrated products show that the feldspars are scarcely altered or affected by decomposition and the biotite no more than partially decomposed. The floor of the basin in which the City of Rocks stands is strewn with the fresh arkosic sands admixed with more or less silt and organic matter. Studies of the more or less disintegrated materials still retained on the outcrop show the same freshness of grain, with only slight evidence of chemical decay along cleavages, crystal boundaries, and fractures. When blocks or slabs of such rock fall from the ledge or monolith, they are at once reduced to sand on striking the ground and because of their disintegrated condition seldom accumulate as fragments (Plate VI, B; Plate VIII). Mineral grains may be readily detached from most of the surface outcrops, and masses of rock, apparently firm, are readily broken to bits with but little effort. Such disintegration of granite is rather characteristic or typical of the weathered granites in desert regions and has been described by Leonard[2] in the arid parts of Arizona, by Blackwelder[3] in the arid parts of the western United States, and by Barton[4] in the almost rainless parts of Egypt. In most places the absence of rock fragments at the base of the ledges or outcrop and the deep massive granular disintegration are the most notable features.

Massive granular disintegration is not the only process active in the Cassia City of Rocks area, but another process, that of case-hardening, is in many places of equal importance, and the two together distinguish this area from most of the others and account for most of the unusual

[1] Ross, C. P., Jour. Geology, Vol. 36, op. cit., also Anderson, A. L., Geology and ore deposits of the Lava Creek District: Idaho Bureau of Mines and Geology Pamphlet 32, pp. 21-25, 1929.
[2] Leonard, R. J., Pedestal rocks resulting from disintegration: Jour. Geology, Vol. 35, 1927, pp. 469-474.
[3] Blackwelder, Eliot, Exfoliation of rock weathering: ibid., Vol. 33, 1925, pp. 793-806.
[4] Barton, D. C., The disintegration of granite in Egypt: ibid., Vol. 24, 1916, pp. 382-393.

and bizarre erosional forms. One generally finds that the outside of the partly decayed rock is more coherent than the inner part. In most cases, but not in all, the exterior has been hardened by the deposition of iron oxides and other minerals under the influence of aridity. It is this indurated or case-hardened shell that serves greatly to strengthen the disintegrated rock and to prevent it from being more rapidly removed by erosion. Excavation proceeds largely beneath such external crusts. The nature and effect of this surface induration is well illustrated in Plate IX, A, where the upper surface of the Giant Toadstool has been strongly hardened and forms an overhanging cap, protecting the massive granular disintegrated rock below. Difference in the two parts of the body is seen not alone from the unequal resistance of both parts to erosion, but in their color as well, the upper surface crust being darker in color and stained considerably with iron oxides. If the indurated shell were to be examined more closely, its surface would be seen to be unequally hollowed or fretted as though having been corroded chemically, whereas the lighter colored rock below would show a smooth surface or only thin exfoliated plates. More detailed examination of the indurated upper surface and of the unindurated rock below is afforded in Plate XI, A, where the contrast in color and the pocking of the surface is especially pronounced. The overhang serves to afford protection for the rock below, but as particles of the disintegrated rock are dislodged or as fragments scale off, undercutting of the upper shell occurs and windows or niches result, which show to good advantage in Plates XI, B, and XII, A. Removal of the disintegrated rock may proceed beneath the case-hardened crust until very unusual and striking forms are produced. The figure illustrated in Plate XII, B, is an example where removal of the disintegrated rock beneath has progressed to an advanced stage, with the production of caverns and tunnels through the rock. Among the forms resulting from such action are large and small hollow boulders, many of them consisting of only a thin shell. Some of them have an opening from the side or bottom through which a man may crawl inside, but many have only small openings on the bottom and also on the side not much larger than a man's head. These may connect with large chambers inside. One of the large hollow boulders is shown in Plate XIII, A. This plate in addition affords a more detailed view of the irregular case-hardened surface and the smooth or slightly scale-like or exfoliated surface on the inside. One of the most curious forms of this kind is the pedestal rock shown in Plate IX, B. This block or case-hardened boulder is balanced on the top of a narrow staff more than 15 feet above the ground.   In the lower side of the boulder is an opening from which the feldspathic sands have been removed, leaving the balanced rock nearly an empty shell.   Other curious features developed as the result of erosion through the case-hardened shell or crust are the numerous depressions or "bath tubs" on the tops of many of the large rock masses such as illustrated in Plate XIII, B.   During heavy rainfalls the running water has cut through the surface shell into the disintegrated rock beneath, rapidly removing the loosened rock and fashioning flat-bottomed depressions or "bath tubs," some of them as much as four feet deep.   These have always a low outlet through which the sands have been washed.   Some of the depressions have a thin veneer of arkosic sand, but most of them have a smooth hard floor, for the wind is also effective in removing the debris.

Speciments of the case-hardened rock were examined microscopically, and these showed the same kind of granular disintegration as the rock below, but generally to lesser degree; and the fractures and cleavages were mainly filled with clay-like products of decomposition, and to lesser extent with iron hydrates.   These products apparently did not result from the decomposition of the minerals in which they were in contact, but have obviously been deposited from solutions which have permeated through the rock.   These materials tend to cement the disintegrated rock particles and prevent the rock from crumbling as in the rock deeper down.

Leonard,[1] who has described pedestal rocks and related erosive features occurring particularly in desert regions, has shown that granular disintegration is the principal process of weathering involved, and results mainly from insolation, although frost action and the wedge-work of salts are no doubt minor contributory factors.   Some of the forms he describes are similar but not so striking as those in the Cassia City of Rocks.   That such disintegration is due mainly to insolation is not borne out by the evidence in Cassia County, inasmuch as the most extensive weathering or disintegration is some distance within the rock mass or beneath the case-hardened shell and especially in the parts of the rock masses most shielded from the rays of the sun.   Blackwelder[2] has shown that daily temperature changes alone are quite inadequate to cause such breakage, and from field and laboratory evidence he concludes that insolation is not an important factor in rock weathering. Similar views are held by Barton,[3] who has studied the weathering of granite in the almost rainless districts of Egypt, where the granitic

---

1 Jour. of Geology, Vol. 35, op. cit., pp. 469-474.
2 Jour. Geology, Vol. 33, pp. 793-806.
3 Jour. Geology, Vol. 24, pp. 382-393.

blocks in that hot and dry region show granular disintegration far below the depths affected by temperature changes and are most weathered in places protected or shielded from the sun's rays. Both Blackwelder and Barton believe that such disintegration may be attributed directly or indirectly to the effect of moisture and present strong evidence in support of their convictions. Blackwelder especially stresses the absorption of water by colloids and the chemical decomposition of minerals, notably by hydration and oxidation of the silicates. Such products are not abundantly developed in the granite waste, but not much decomposition is necessary to provide for the expansive force needed to disrupt the rock or loosen it. The expansive force induced by the crystallization of secondary minerals of larger volume along cleavage cracks of crystals and between mineral grains is all that is necessary.

The principal process of weathering involved in the production of the particular erosional forms in the Cassia region seem to be chemical and the work has been performed by solutions which penetrate slowly along the cleavage cracks in crystals and between mineral grains and there induce the formation of new minerals of larger volume. Although the degree of decomposition appears to be slight, the amount of hydration and other chemical changes is sufficient through increased volumes to disrupt or disintegrate the rock into the condition now found existing in the region. These slight changes may be noted throughout the rock in thin sections, though not always apparent to the unaided eye. During the dry season the sun quickly dessicates the outer side of the rock masses after the occasional rains, and that moisture which has penetrated far into the rock through the minute fractures, and that which remains on the shady undersides continue the chemical work for a much longer time and promote greater disintegration. As the outer or near outer surface of the rock masses tend to become dessicated, capillary action draws water from the interior of the rocks and with them dissolved salts and colloids, and as this capillary water in turn evaporates at or near the surface it deposits its borne substances, thus cementing or closing the fractures in the rock and thereby strengthening the disintegrated rock by the process of case-hardening. Granular disintegration may continue beneath the case-hardened shell as long as moisture is retained, and if the products of disintegration are free to fall or blow away, overhanging ledges, niches, and hollow boulders result. Frost action or alternate freezing and thawing during the spring and autumn may aid in rock breakage or disintegration, but most of the granular disintegration is probably the result of hydration and other processes of chemical decomposition. Some exfoliation results from this action, as discussed by Blackwelder, but the amount as shown in the several plates is usually not conspicuous. Such coarse-grained rocks seemingly do not lend themselves to exfoliation processes.

Similar cavernous and pitted surfaces have been recently described by Blackwelder,[1] who ascribes mechanical disintegration induced by expansive chemical changes as probably the most important single factor involved and the wind its chief, but by no means only, aid in clearing the cavities. He also states that the same process-complex is responsible for such features as mushroom rocks, perched boulders, and other picturesque details of the scenery of a dry region. The writer's observations confirm those of Blackwelder.

At first glance the areal arrangement of the numerous isolated outcrops of granite appears chaotic, but it is controlled by the jointing of the original igneous mass as discussed in another section. Joint planes have greatly facilitated the weathering of the granite and have permitted the solutions to penetrate to great depths and there to carry on their work of alteration and attendant disintegration. At the same time these planes have tended to direct and serve as channels for the flow of surface waters and, widened by this action, have helped to isolate the individual joint blocks from one another and form the high isolated monoliths, turrets, spires, etc. As undermining of the sides of the projecting blocks of massive rock by the widening of the joint planes by running water and by the gravity removal of the disintegrated rock from beneath the case-hardened roof has occurred, large blocks or slabs have broken loose and have produced such forms usually ascribed to insolation. The Cassia City of Rocks is apparently the peculiar result of a complex set of factors involving deep granular disintegration and case-hardening under especially favorable climatic conditions, together with especially favorable structural features, not alone the widely spaced sets of vertical and horizontal joints, but also the protective quartzite capping on the upper side of the basin and the flanking ridge on the lower side which has maintained a proper balance between erosion and weathering by not permitting too rapid removal of the waste products such as would have occurred had erosion not been retarded by the more resistant rocks.

EXTRUSIVE ROCKS

TERTIARY SERIES

Considerable discussion has already been given the lavas interbedded with and capping the sedimentary strata in the upper division

[1] Blackwelder, Eliot, Cavernous rock surfaces of the desert, Am. Jour. Sci., Vol. 17, 1929, pp. 393-99.

of the Payette or Salt Lake formations. In areas described by Mansfield[1] in southeast Idaho the lavas are dominantly rhyolites, subordinately basalts and latites. In western Cassia County, Piper[2] has classed the lava as rhyolite, although Larsen in his microscopic analyses of the rocks lists them as quartz-tridymite latites. Sufficient specimens were collected from the various flows over eastern Cassia County to show that the flows invariably contain enough plagioclase to class them as latites rather than as rhyolites. Basalt is locally intercalated in the series.

QUARTZ LATITE

Distribution: The same flows of lava described in the Goose Creek area extend upon the lower western flank of South Mountain and well along the west base of the Albion Range. The flows are well exposed in the canyons south of Oakley and along the fault scarp which has brought the edges of the lavas above the alluvium-floored valley east and northeast of Oakley. As the flows have been unequally eroded, it is difficult to determine their exact number and precise thickness. West of South Mountain two flows appear in the upper canyon walls, but in some places only one remains, and near the Utah line it, too, has been mainly eroded. The lower flow is probably not much over 50 feet thick and the upper not over 115 feet. In the scarp east of Oakley four flows appear and present a massive outcrop nearly 500 feet high (Plate XV, B). Measurements of the individual flows were made and these show the top flow to be 88 feet thick, the second 132 feet, the third 145 feet, and the lowest 50+ feet.

Some patches of lava are scattered in the Albion Range. The most extensive adjoin the town of Albion and extend several miles to the north. That adjoining Albion is a large landslide block capped by a flow nearly 150 feet thick. Two small remnants also lie near the south end of the City of Rocks area, one in the floor of the basin on the granite and the other high up on a quartzite knoll on the west rim of the basin. Remnants of a flow also lie on the flank of the range west of the Elba Basin.

The largest continuous exposure is in the Malta Range. In the north section of the range the lava surface is but little dissected or interrupted by gullies. In the higher middle and southern sections, where dissection has been much more extensive, the lava forms tilted mesa-like caps between gullies. As much as 500 feet of lavas appear in the top of the steep eastward-facing scarp (Plate XIV, B), but again

---

[1] Prof. Paper 152, op. cit., pp. 116–129, also Bull. 713, p. 6.
[2] Idaho Bureau of Mines and Geology, Bull. 6, op. cit., pp. 27, 30.

the number and thickness of the flows is variable. Near the top of the range west of Idahome three flows are certain, the uppermost at least 300 feet thick, and the one below 100 feet thick. These are separated by very little sedimentary material. Numerous detached landslide blocks add to the apparent number of flows and these also conceal the lower parts of the series and it is therefore difficult or impossible to obtain the exact number of flows in the range. Farther south erosion has removed all but the lower thick flow. This one is about 300 feet thick.

Across Raft River Valley the upper latite flows have been stripped off and only one or two well down in the series remain, including a flow of basalt. A black, glassy latite or rhyolite flow has wide distribution. This flow is less than 50 feet thick, but it serves to conceal most of the underlying sedimentary strata.

Description: Differences in composition between the flows are no greater than variations in a given flow at separated points. Most striking differences are physical rather than chemical. Thick flows have usually a black basal porphyritic obsidian zone from a few feet to as much as 65 feet thick, above which the latite is often exceedingly vesicular. The vesicular portion in turn is usually overlain by a denser portion, often of great thickness, pinkish or purplish in its lower part and bluish-gray above, most of it with a pronounced thin platy parting or sheeting, consisting of feldspar and frequently quartz phenocrysts in a felsitic groundmass. Usually the bluish-gray facies extends to the top of the flow, but in some it grades upward into a thin black porphyritic obsidian, not more than 15 feet thick. This sequence is not everywhere the same and some of the thinner beds, as east of Raft River, consist largely of the black glassy facies inclosing phenocrysts of white feldspar and usually quartz. Reddish tints may also be lacking in some and the bluish-gray color in others. In general the flows in the Malta Range have only the obsidian and the grayish felsitic facies and those along the west border of the district the obsidian and the pinkish colored rock. Portions of the flows have well developed spherulitic structures in both the obsidian and the felsitic facies. These spherulites commonly measure one-half to an inch in diameter, but spherulites which measure five inches in diameter also occur. Some of the flows in the Malta Range possess an exceptionally well defined columnar jointing and display remarkable columnar forms (Plate XIV, A). Within a short distance the polagonal jointing may disappear and give way instead to the normally thinly plated rock. Some of the flows have a distinct fluidal banding, but this is usually not so pronounced in the thicker flows.

Microscopic examination also indicates that differences in the flows are mainly physical and have depended wholly on the rate of cooling. The glassy and felsitic facies are equally porphyritic and contain the same phenocrysts. Differences exist in the ground-mass, glassy in the obsidian facies and usually partially glassy to microcrystalline in the felsitic zones. Differences in the rate of cooling from the outside to the inside of the flows are also reflected in the presence of much tridymite in the outer parts and of quartz well within. From the same flow specimens might be classed as vitreous latites, tridymite and quartz latites, augite-quartz latites, and augite-tridymite latites. Flows composed of black obsidian alone usually have the same kinds and the same propostions of phenocrysts as the glassy facies of the thicker flows and this relation suggests that these, too, are latites and not rhyolites. In general, the phenocrysts consist chiefly of plagioclase (acid andesine) and lesser augite and often quartz. These normally comprise from 20 to 25 per cent of the rock.

Latite collected near the south end of the Malta Range and typical of the thick grayish-colored facies above the lower obsidian zone may be considered as representative of the lava in that end of the range. It has scattered plagioclase laths in a grayish aphanitic groundmass, which in thin section shows andesine, occasional resorbed quartz and augite as phenocrysts, in which the andesine makes up about 20 per cent of the mass. These are embedded in a finely crystalline groundmass of tiny orthoclase, oligoclase, augite, and quartz crystals. Apatite, zircon, and magnetite are accessories. Specimens from the flows west of Idahome are similar, but phenocrysts are slightly more abundant, especially those of augite. The phenocrysts comprise about 25 per cent of the rock. A section of the obsidian facies shows the same phenocrysts but embedded in a glassy groundmass with porous streaks partially filled with tridymite.

Latite used as building stone by the Albion State Normal School and quarried at the edge of the town of Albion has a pleasing purplish-red color. Most of the flow is highly vesicular, even the obsidian base, and most of it has a thin platy parting, but a zone five or six feet thick well toward the center permits quarrying. This rock is porphyritic, also slightly vesicular or porous adjacent to phenocrysts. Andesine is the only mineral occurring as phenocrysts and comprises about 20 per cent of the rock. It is embedded in a gray to reddish-brown mottled groundmass, in part finely crystalline but mainly glassy with pronounced fluidal banding. Numerous porous streaks occur in the groundmass and most of them are partially filled with aggregates of tridymite. This mineral constitutes about 20 per cent of the rock. Accessories include augite, zircon, apatite, and magnetite. Elsewhere the reddish or pinkish-tinted latites always carry tridymite and the bluish-gray rock always carries quartz. Apparently then, color of the rock is a factor of its temperature of solidification. A lavender-tinted latite from a small remnant of a flow in the City of Rocks area has laths and rounded crystals of acid andesine and occasionally partially resorbed quartz grains in a minutely vesicular groundmass of glass, alternating with streaks or porous zones of tridymite, less commonly associated with minute crystals of plagioclase, orthoclase, and quartz. A similar reddish-purple colored latite above an obsidian base forms the flow west of the Elba Basin. It has very few phenocrysts and its main mass is glass dusted with hematite and limonite and holding numerous porous or semi-porous streaks largely filled with tridymite. In both places the tridymite may comprise about one-fourth of the rock.

Each of the four flows in the scarp east of Oakley were also examined. These show no appreciable differences in color, texture, and composition. Each flow is reddish tinted, varying from pale lavender-gray to distinctly reddish, and each is more or less vesicular, and shows flow banding. All, however, do not contain the black glassy base. Phenocrysts are acid andesine laths and augite in a glassy to partially crystalline groundmass; in either case, with porous streaks and layers containing much tridymite. Accessories are the same as in other flows. Tridymite also occurs in great abundance in lava in the adjoining Goose Creek district, and its rocks are no different from those near Oakley or in the Albion Range.

Rock on the east side of the district shows little microscopic change from that on the west side. A purplish or lavender-tinted rock from a small fault scarp southwest of Heglar has scattered plagioclase laths, a few augites and rounded quartz crystals, in a partially glassy groundmass with porous and crystalline streaks rich in tridymite and also carrying a little orthoclase and plagioclase. Some of the black glassy lava so widespread east of Raft River Valley is locally rich in phenocrysts composed of andesine laths, many of them broken, and also broken quartz grains in a brownish glassy groundmass with pronounced fluidal structure. The rock has no microlites, but has numerous small grains of magnetite.

BASALT

Distribution: A basalt flow, which is intercalated with the sedimentary strata in the Sublett region, presents an outcrop nearly 40 feet high for about two miles along the Sublett Creek road, east of Sublett post office. It was not feasible to map this flow separately from the others, as its presence is almost at once obscured by the sedimentary strata above. A small remnant of basalt also occurs at the extreme north end of the Malta Range, lying above the thick latite flow.

Description: The Sublett basalt is a dull grayish black rock of faintly purplish cast, highly vesicular with vesicles ranging from microscopic size to an inch in diameter, usually partly filled with calcite and zeolites. Microscopic examination reveals that the rock is an olivine-rich basalt composed of olivine (15%), titaniferous augite (25%), labradorite (57%), ilmenite (3%), and a little magnetite and apatite. It is faintly porphyritic, for some of the plagioclase laths reach somewhat larger sizes than others, but its texture is dominantly ophitic, with the augite in the interstices between the plagioclase laths or enclosing them poikilitically. Much of the olivine has altered to a reddish brown mineral resembling iddingsite.

The remnant flow at the north end of the Malta Range has a dark gray color with the same purplish cast, but the rock is dense and so coarse-grained that plagioclase laths may be seen without the microscope. This basalt is medium grained and is composed also of labradorite (60%), augite (20%), and olivine (20%), and holds magnetite and apatite as accessories. Some of the plagioclase laths might be regarded as phenocrysts, but they are not much larger than those in the groundmass. The texture is ophitic and the augite tends to hold several of the plagioclases at once in poikilitic fashion. Most of the olivine shows varying stages of alteration to iddingsite and it is the reddish brown color of this mineral that gives the rock its purplish cast in the hand specimen.

CORRELATION AND AGE

Correlation of the lava series with the Tertiary flows in southwest and southeast Idaho has already been suggested. As the age of the flows depends on the age of the intercalated sedimentary strata it is certain that most of the latite belongs well up in the Upper Miocene, but the age of the uppermost flow is none too well established. In the upper Snake River Plains area Kirkham[1] assigned the series to the

---

[1] Pamphlet 19, op. cit., pp. 31-33.

Pliocene and calls it the late Tertiary lava series.' This assignment was made because he considered the Salt Lake formation in which these lavas occur to have a Pliocene age and also because the lavas are continuous with some of the flows of the Yellowstone National Park which are regarded as Pliocene. Mansfield[1] has likewise regarded them as Pliocene for essentially the same reasons. More recently Kirkham[2] has shown the rhyolites in the Boise region to be underlain by sediments yielding a Miocene flora and, overlain by the Idaho beds containing a Pliocene flora. He believes therefore the rhyolite to be of Upper Miocene or Lower Pliocene age. It is possible that the extrusions continued from the Miocene into the Pliocene and that the topmost flow is Pliocene. Until more data are available it is not advisable to establish any fixed age for the flows.

QUATERNARY SERIES

SNAKE RIVER BASALT

Distribution: Basalt continuous with that of the Snake River Plains area extends well into the northern part of the district. One area partially encircles the northern ends of the Albion and Malta ranges and embays for six miles or more into the Raft River Valley. The source of the basalt is local and from low lava domes and vents north and south of Cotterel (Plate XIV, B). From these vents the basalt spread westward around the two ranges to Declo, southward and eastward into Raft River Valley, and northward toward the central Snake River field. There is another area of basalt in the extreme northwest corner of the district four or five miles west and southwest of Burley, where the lava has again spread in all directions from a low dome or flat cone. A low basalt dome also rises above the alluvium of Goose Creek Valley about nine-miles south of Burley near the west base of the Albion Range. Much of the area mapped as alluvium between the basalt areas is underlain by flows.

Description: The basalt flows in most places have not been eroded and for this reason can seldom be observed below the surface except in road cuts or where Raft River flows around the outer margin of the flow in Raft River Valley. Thicknesses of the surface basalts are not entirely known, but probably range from 20 feet or less on the flanks of the domes to 100 feet or more at the margins.

The flows have a dark gray to nearly black color and show little or no signs of alteration or weathering. The upper surface is in general

1 Prof. Paper 152, op. cit., p. 129.
2 Jour. Geology, Vol. 38, op. cit., p. 652, also Jour. Geology, Vol. 39, op. cit., pp. 193-239.

ropy and is essentially in its original condition. Most of the flows are highly vesicular and it is difficult anywhere to find dense rock. Vesicles range from microscopic to a foot or more in diameter and are especially numerous in the upper 10 feet of the flows. Even the denser central part contains numerous minute vesicles and occasional larger ones. Flows are also broken by widely spaced joints which produce rude vertical polagonal columns of large size. Most flows have rock of a pronounced crystallinity, ranging from fine to medium grain, and it is not difficult to detect plagioclase laths and small grains of olivine without a lens. Olivine is comparatively abundant in most of the flows and frequently gives a greenish cast to the rock.

The flows show everywhere a monotonous similarity of composition and texture whose differences from flow to flow are no more than within the same flow. Differences depend mainly on the rate of cooling. The texture is typically subophitic, and the rock is usually free from glass. Most of the rock is slightly porphyritic and contains a few scattered phenocrysts of plagioclase, olivine, and augite. These are but little larger than the crystals in the groundmass. Labradorite laths lie in all directions and between them occur the augite grains and most of the ilmenite laths. In general the labradorite constitutes about 55% of the rock, augite 30%, olivine 10%, and ilmenite about 5%. The augite is the titaniferous variety and has a brownish-purplish color in thin section. The ilmenite tends to form lath shapes and penetrates both the plagioclase and the augite. Accessories are apatite and magnetite.

Age: When the basalt was extruded, the topography was essentially as it is today. Surface rock shows no weathering, nor have the cracks and irregularities of the surface of the flows been everywhere filled with alluvium or aeolian debris. Because of the freshness of the basalt flows, workers have in general assigned the extrusions to the Quaternary, most of it to the early Pleistocene, but some, as in the Craters of the Moon National Monument directly north of Cassia County and on the opposite side of the Snake River Plains, to recent or historical times. That in Cassia County is surely Pleistocene as it post-dates several of the Pleistocene erosional cycles as well as follows after Pleistocene deformation.

STRUCTURAL GEOLOGY
GENERAL FEATURES

Eastern Cassia County possesses structural features of much more than ordinary interest, for in the area there is found the marked folding and great low-angle overthrust faults so characteristic of the Northern

Rocky Mountain region; also the young block faulting of the Basin Range country; and finally the great cross fold involved in the Snake River downwarp. Each of the structural features is clearly the product of separate, unrelated, diastrophic movements. In effect, the Basin-Range structure is superimposed on the earlier Rocky Mountain structures, and the Snake River downwarp is superimposed on the other two.

The structure evolved during the early diastrophism is highly complex and much the same as the Rocky Mountain structure so typically developed in southeastern Idaho. Most of the individual folds are developed as parts of anticlinoria or synclinoria of large magnitude. They are also mostly asymmetrical and overturned or over-steepened to the east, as though acted on or produced by forces directed from a westerly or southwesterly direction. The great overthrust faults, which formed near the end of the period of folding, are the most interesting of the structural features. Their general trend conforms with that of the folds, and like the folds they were produced by great tangential pressures acting mainly from a westerly direction. The faults are of the low-angled kind and have nearly horizontal planes, warped into gentle folds or tilted as a result of later deformation. Movement along the fault planes has been great and is to be measured in dozens of miles. Recognition of these great overthrusts is largely from "windows" of younger strata beneath the older rocks. Transverse faults of large magnitude have interrupted or displaced both the folds and the faults.

Each of the ranges disclosing the older rocks differ among themselves in the direction of trend lines and in the degree and magnitude of folding and faulting. Yet each is closely bound with the other in the character, origin, and time of structural development and a common basis for structural inferences therefore exists for all the areas of older rocks, even though widely separated by an intervening cover of younger formations. Interruption of the continuity of trend lines has in most cases been due to transverse faults of great magnitude, and, although the prevailing trend lines are from north to northwest, not a few trends normal to these have resulted from such faults.

The young block faulting has broken and displaced the early Rocky Mountain structures and has brought some of the ranges into their present topographic forms. These faults are of the normal kind and have in part produced tilted block mountains very similar to the young block mountains in the Harney Lake region of Oregon, the reputed youngest block ranges in the Basin-Range province. The faults trend in a northerly direction for the most part and are not in any way controlled or directed by the earlier trend lines.

The Snake River downwarp, which has been entirely instrumental in forming the setting of the Snake River Plains, is a great trough or basin caused by subsidence. Its position is directly over that of the earlier structures but in a general direction at right angles to these lines. Thus the Rocky Mountain structure and the Basin-Range faults approach its margin and then plunge downward to disappear beneath the lava-flooded floor of the downwarped area, to reappear again, at least the Rocky Mountain structures, on the north side of the Snake River Plains.

Time allowed for general studies within the district did not permit recognition or mapping of all the minor folds and faults, but was sufficient to determine the major structures as outlined above. Many of the smaller structures were especially difficult to decipher because of the great thickness of some of the formations, the small number of reliable horizon markers, and the abundance of surface debris. But the lack of an adequate base map on which minor features could be accurately placed was even less favorable to detailed observation and detailed mapping. All available data have been taken into consideration in the interpretation and illustration of structural features, but some of the sections must be regarded as essentially diagrammatic. In many places structure sections extend to depths far below the limits of observation. Question marks have been inserted in places to indicate that in these places direct evidence is lacking and the attitude of the beds has been sketched on purely inferential grounds. Many minor faults have been omitted from the map and structure sections.

The easily recognized unconformity at the base of the Tertiary series is one of the most useful means available in fixing limits to the possible age of structural disturbances. Obviously, any fault or fold in the older rocks which is truncated by this unconformity is of earlier date than the erosion period which ended with the deposition of the Miocene (?) sedimentary strata and volcanic rocks. Faults and folds which involve the Tertiary rocks are obviously of younger age, and, in turn, folds which affect the structures developed in the Tertiary rocks are still younger.

FOLDS

Folding has affected all the pre-Tertiary strata to marked degree, although the character of the formations has mainly determined the complexity of the folding. Strata of competent nature have for the most part been arched or compressed into simple folds, whereas incompetent beds have been more intensely crumpled than the others. Folds in the pre-Cambrian strata are therefore most simple in outline, for the very competent quartzites, which comprise most of the series, have yielded less readily to the great compressive stresses than the younger, less competent Paleozoic rocks. In the latter, folds are more numerous and tend to develop as parts of synclinoria and anticlinoria rather than as broad synclines or anticlines. In beds of most incompetent nature the folding is especially intricate. Unfortunately, folds in such strata could not be entirely worked out in the field and the sections are in consequence much simplified in diagram. Much of the faulting has to be discussed in connection with the folds, but independent consideration is given to the major overthrusts and to certain other faults.

ALBION ANTICLINE

Folding in the Albion Range is of the simple anticlinal kind, except locally in blocks of younger strata appearing as "windows" beneath the pre-Cambrian rocks. The topography in general conforms so closely with the structure that the range might be regarded as truly anticlinal. Its structure, however, is not as simple as the anticlinal character might suggest, for, as inferred above and discussed in another section, the anticline is a block of overthrust pre-Cambrian strata and is underlain by Paleozoic strata much more intricately deformed.

Mention has already been made of the two-segment character of the range, the southern domed-shaped segment culminating in Cache Peak and the northern domical section culminating in Mount Harrison. These two sections of the range are controlled by the structure of the rocks and actually represent two domes which together form the anticlinal backbone of the range. Each has nearly perfect enclosure, but the area between them is much disturbed by faulting and is in addition interrupted by a "window" of Carboniferous strata.

As already stated the southern segment is composed mainly of the lower part of the Harrison series with an intrusive core of granitic rock (as shown in structure sections D-D', E-E', and F-F'). The western slope of the range is nearly a dip slope in which the strata are tilted from 8 to 20 degrees, locally somewhat steeper. Strata on the east side of the anticline, as represented mainly by the border of hogback ridges, are inclined from 20 to 25 degrees. Thus the anticline or dome is slightly oversteepened on the east. (Structure section D-D'.) Intrusion has apparently affected the simple outline of the fold but very little. The structural axis trends a little east of north.

The structure of the northern section lacks the apparent simplicity of that on the south only because closure has not been as effective and because erosion has carved more deeply into the southern and north-eastern parts of the dome. Faulting has also been a big factor in causing its distortion. This dome is much wider than the one farther south and the dips are slightly steeper except on the east (structure sections A-A', B-B'). The long axis of the dome trends slightly east of north and lies well toward the eastern margin of the range, as is so well shown in structure section B-B'. For this reason all but the lower beds have been eroded from the east side of the dome, whereas nearly the complete pre-Cambrian section is exposed on the west side. North-east of Mount Harrison the entire crest and east limb of the anticline have been faulted and apparently deeply eroded as well as later concealed beneath younger strata. Only a part of the west limb remains. North of the town of Albion and extending to the Snake River Plains this limb has been so broken by faults that its relation to the fold is not readily apparent. Nevertheless the dips are persistently to the north-west, and when allowances are duly made for the offsets by numerous transverse faults, its position on the anticline is obvious. North of Mount Harrison the structural trend swings increasingly to the north-east and finally approaches northeast by east. At the same time the range plunges northward until finally it passes beneath the Snake River Plains.

Carboniferous rocks in the window at the north end of the range and also in the one between the two domes in the mid-point of the range are much more intricately folded. Structure in the northern window was not ascertained as exposures were too limited to permit adequate interpretation. The window in the central part of the range, however, is much larger and permits more satisfactory study. The Pennsylvanian strata as shown in structure section C-C' have apparently been compressed into a series of anticlines and synclines, of which two anticlines accompanied by two synclines appear in the window. These minor folds are unsymmetrical and have for the most part steeper dips on the east flanks, a relation which suggests pressure from the west as the cause of the folding. A short distance to the north from where the section was taken the structure has been further complicated by faulting and the folds entirely offset. Not enough of the strata is in view, however, to provide for general interpretation.

BLACK PINE FOLDS

Folds in the Black Pine Range constitute a series of anticlines and synclines of considerable magnitude. Faulting of both thrust and transverse kinds has destroyed the continuity of the folds, and, as a result, the range structurally as well as topographically is divisible into three main sections. Each displays a difference in trend and structural character which makes individual treatment of each section necessary.

The southern section of the range contains two anticlines and two synclines, as shown in structure section N-N′, in which the anticline in the center of the range has the greatest magnitude. The folds are in the main gentle and the dips seldom exceed 20 degrees, their average dip being about 15 degrees. The folds are unsymmetrical. Their general trend is northward, but inasmuch as the range plunges southward the structure tends to show partial closure on the south, and the beds on the west side of the range trend to the northwest at angles ranging from north 30° west to north 45° west and beds on the east side of the range strike about north 20° east. Anticlinal and synclinal axes show similar trends. The axis of the main anticline follows closely the ridge between Rice Canyon and Formation Canyon. The chief syncline lies to the east and approximates the course of Black Pine Canyon. The anticline on the west side of the range is only a minor one and the syncline between is also on a small scale. This structure is ended abruptly on the north by a transverse fault which is followed by the general course of Kelsaw Canyon and which is continued in an easterly direction entirely across the range. Strata forming the folds are composed of the Brazer formation. Along the southeast side of the range these beds apparently pass over westerly tilted beds of the Lower Wells formation and from their position suggest an overthrust sole. Not enough of the underlying block is exposed to work out its general structure, but its dip ranges from 20° to 35° to the west and its strike from northwest to northeast at angles ranging up to 20° on both sides of north. This block apparently represents the west limb of an anticline.

Folding in the middle section of the range (structure sections M-M′ and L-L′) has been greatly complicated by faults of normal and thrust character, and the strata more nearly resembles tilted blocks. Unlike the general northerly trend of the southern segment of the range, the strata have a persistent strike to the northwest of from 15° to 20°, except on the west side, where the structure is a probable continuation of that of the north segment of the range. Along the eastern margin of the middle segment the Wells formation has been compressed into an anticline with steepest dips on the west side. This anticline has been sliced on the west by a fault which trends northwest from the head of Mill Creek Gulch. The block on the west side of the fault has inclined strata whose dips are about 30° west and strikes about north 15° west. In turn this block has been sliced by a strike fault and again the strata on the west appear in a tilted block, whose general trend is from 15° to 20° west of north, and dip from 20° to 30° west. This block forms the highest part of the range. Far down on its west flank there appears an over-riding block whose beds strike northeast and dip in general to the northwest, but much crumpling and faulting has so disturbed them that the exact structure was not deciphered. Normal faulting has also occurred on that side.

Folding in the north segment of the range has been less complicated by faulting. In this segment structural trends are at right angles to those in the middle segment and the beds and structural axes strike persistently from 30° to 45° to the northeast. This abrupt interruption of structural lines is again to be explained by faulting. Near the southern end of the segment the strata form an anticline oversteepened on the south side (structure section O-O'). Most of the segment, however, has a synclinal structure and the highest part of the segment is near the axis of the syncline. This syncline also has steeper dips on the northwest flanks. A minor anticline at the north tip of the range is nearly concealed by younger strata. Near the southwest edge of the segment there appears a window of younger Wells formation whose general strike and dip differs somewhat from that of the overlying strata. It appears to form the west flank of a syncline.

Tertiary strata on the flanks of the range dip away from all sides at an angle of about 15°. Apparently these strata once covered the entire range, but were subsequently removed when the range was arched into its present position.

SUBLETT FOLDS

Structures in the Sublett Range are less satisfactorily deciphered. Large overthrust and normal faults have broken the folds and have made interpretation of the folding in many places impossible until close detailed work can be done. Unfortunately, only the western part of the range was studied in the reconnaissance and it is only by work over the entire range that the structures can be entirely interpreted.

The main structural feature is an overthrust fault that has brought the Brazer formation above the Wells-Phosphoria in the northwest part of the range. Strata both above and below the fault plane are folded.

Folding in the overthrust block of the Brazer formation is comparatively simple, as illustrated in structure sections G-G' and H-H', and consists of three anticlines and three synclines of minor magnitude, whose axial trends are north 20° west. These folds are slightly steeper on their east flanks than on their west, but the steepest dips rarely exceed 20°, and in some places they nearly approach horizontality. Steepness of dips appears to increase southeastward along the range and reaches a maximum about three miles southeast of Cedar Peak, where the fold at the crest of the ridge is nearly closed and over-turned. The block with its series of folds appears to have an eastward tilt, and also a northward inclination that carries it beneath the Snake River Plains.

Folds in the underlying Wells-Phosphoria formations are much more complex, especially as these have been more broken by faults. These folds also consist of anticlines and synclines whose general trend north of Sublett Creek is about north 20° west, essentially parallel to the trend of the beds and folds in the Brazer block (structure sections G-G', H-H', I-I'). On the ridge east of Sublett the strata form an asymmetrical anticline with the steeper limb on the northwest side (structure section J-J'). Relations along Van Camp Creek are concealed by Tertiary strata and it is not possible to relate the anticline to structures on the east side of Lake Fork. Several minor folds may be concealed, as suggested farther north, but it is more probable that the strata have been broken by a strike fault. East of Lake Fork the strata have a variable inclination to the northeast ranging from 10° to 50°. Dips are most gentle at the top of the ridge. In the region west of Heglar Ranger Station, several minor anticlines and synclines are inferred, and these probably pass beneath the Tertiary rocks in Lake Fork drainage area. Folds or trend lines in the detached parts of the range bordering Raft River Valley and the Snake River Plains conform with the folds or trend lines of the main mountain mass. In only one outlier, that at the extreme north end of the district, do the strata trend otherwise. In general, the Phosphoria formation appears on the flanks of the anticlines, but faulting has greatly obscured its relations and true positions.

A transverse fault along lower Sublett Creek has caused offset of folds and trend lines. Adjoining the fault on the south the trend is approximately at right angles to that on the north side, and is about north 40° east. Between Sublett Creek and Cold Spring Canyon the strata are arched into an anticline whose southeastern flank has an average dip of 20° and its northern flank from 10° to 25°. Apparently the middle part of the Wells formation is involved in the folding, and the lower area of Phosphoria rocks on the west owe their position to faulting. A transverse fault is also followed by Cold Spring Canyon. South the strata maintain the same northeasterly trend and dip south-east, but south of Pine Creek the tilted inclination gives way to gentle folds and at least one syncline and one anticline are exposed. These relations are illustrated in structure section K-K'.

Tertiary strata show gentle inclinations and in general dip away from the flanks of the range with angles as high as 15°. It is probable that at one time the Tertiary rocks actually extended over the mountainous mass and have been largely stripped off, except on the lower flanks, by subsequent up-arching and erosion of the range. Such strata as now remain are mainly in synclinal depressions or remain as remnants in the older pre-Upper Miocene valleys, as yet incompletely reexcavated.

FAULTS

OVERTHRUST FAULTS

ALBION OVERTHRUST

Greatest of the overthrust faults is the one in the Albion Range, which has carried a block of pre-Cambrian strata over Carboniferous rocks (structure sections A-A', B-B', C-C'). It is proposed that this fault be called the Albion Overthrust. Presence of the overthrust is inferred in several places within the range where "windows" of the younger rock have been exposed through erosion of the overlying block. The pre-Cambrian strata may be seen directly on the Paleozoic rocks. In the Elba Basin and on the divide to the west the contact may be readily followed for several miles in the mountain wall (Plate XV, A). The divide itself is still capped by schists of the middle division of the Harrison series, but a short distance below strata of the Wells formation appear on both sides of the capping. The trace of the fault plane about both windows is irregular, although inclined to be oval or circular.

The fault plane has apparently been warped like the plane of the Bannock overthrust.[1] In the area west of Elba the fault plane has an inclination to the east and passes beneath pre-Cambrian strata before reaching the edge of the basin or the east edge of the range. On the west it appears high on the flanks of the range and has probably been cut off or dropped below the zone of observation by younger normal faults. Relations are similar at the north end of the range.

The magnitude of the displacement cannot surely be known, but must be very great. The stratigraphic throw represented by middle strata of the pre-Cambrian series on Pennsylvanian and possibly Permian rock involves a displacement of at least 20,000 feet, and more probably more than twice that figure. This, however, gives little clue to

---

[1] Mansfield, G. R., Prof. Paper 152, op. cit., pp. 150-159.

the magnitude of the horizontal displacement. Younger strata effectively conceal the formations involved in the faulting east of the Albion Range, but fortunately warping along an east-west axis in the Raft River Range has brought the older rocks again to view and, in the lower parts of the range, the fault plane may be readily traced from near the south end of the Albion Range to near the east end of the Raft River Range. Here the moderately warped fault plane may be seen for 17 miles from east to west before younger normal faults drop it below the zone of observation. Throughout this distance the overriding block of pre-Cambrian strata is underlain by folded Paleozoic rocks and the lip of the thrust block lies not far from the south end of the Black Pine Range and upon the same folded series of strata as outcrop in the south end of the Black Pine Range. It is not improbable that the overthrust block actually overrode the Black Pine Range, and this suggestion has partial confirmation in the finding of quartzite and schist float from the pre-Cambrian series on the flanks of the Black Pine Range. Warping of the fault plane may have carried the upper block to such elevations that it could not survive subsequent erosion, and this possibility is suggested by the rapid steepening of the fault plane at the east end of the Raft River Range. Actual displacement along the thrust plane probably exceeds a score of miles and possibly exceeds two score. Its magnitude is probably equal to that of the Bannock overthrust farther east to which Mansfield credits a movement of 35 miles or more. Both are very similar in that both are low-angle overthrusts approaching horizontality, in which the fault plane has subsequently been moderately warped and faulted so that the overlying block is cut through here and there by erosion, thus exposing the younger rocks beneath. Mansfield has shown that, starting in northeastern Utah, the sinuous course of the Bannock overthrust may be traced northward and northwestward with interruptions for about 270 miles. It is even possible that the Albion overthrust is merely a more westerly extension of the Bannock thrust plane, or a closely associated break.

BLACK PINE OVERTHRUSTS

Overthrusts of more moderate magnitude and perhaps subsidiary to the Albion overthrust appear in the Black Pine Range. One of these is inferred in the south segment of the range, where the Brazer formation appears above the Lower Wells on the lower southeast slope of the range. Displacement stratigraphically is perhaps 2,000 or 3,000 feet, but the horizontal or actual movement is probably much greater (structure section N-N').

Overthrusting is also the main structural feature in the remainder of the range, but whether the overthrust is a continuation of that described above is not certain, especially as the stratigraphic horizons involved appear to be different. The overthrust block appears along the west side of the middle section of the range (structure sections L-L' and M-M') and includes the entire northern section (structure section O-O'). It partially surrounds the high middle section of the range on the west and north. Apparently the fault plane is fairly steep where the upper block lies in contact with the lower, but the general character is that of a low-angle thrust whose plane has been warped or folded. Evidence of the general low angle is afforded from a window of the younger strata at the west base of the range. Stratigraphic displacement is not great, as the movement has only brought beds of the Lower Wells upon the middle and upper divisions, but this is not a reliable basis for ascertaining the actual movement along the fault plane.

SUBLETT OVERTHRUST

Considerable attention has already been given to the low-angle overthrust in the Sublett Range, which has brought a sole of the Brazer strata over Wells and Phosphoria beds. The fault trace may be readily observed from where it enters the district from Power County along a headward tributary of Sublett Creek, and as it crosses the low saddle into the head of North Heglar Canyon. Its trace then curves to the west and northwest in the upper end of Long Canyon, and then crosses the divide into the Lake Fork drainage. It then follows the head of the Lake Fork drainage for some distance to the northwest and passes over the low divide into South Heglar Canyon. Its trace is more or less closely followed by South Heglar Canyon to its disappearance beneath the Tertiary strata not far to the northwest. The fault thus has a very sinuous course. South Heglar Canyon has cut through the upper block, and in so doing has left several isolated caps or outliers on the west side of the canyon, resting on the Wells-Phosphoria formation. But west and south of the trace as outlined the strata belong essentially to the middle and upper Wells and to the Phosphoria formations from which the overthrust Brazer has been almost entirely removed by erosion because of the apparent uparching of the fault plane to the west.

Like the thrusts already described the fault plane has been slightly folded and appears to have a gentle inclination or dip to the east that causes the underlying block soon to plunge from sight (structure sections G-G′, H-H′ and I-I′). Displacement along the fault plane is conjectural, but is presumably great and is probably measured in thousands of feet, possibly in miles. The stratigraphic displacement must reach three or four thousand feet. It is possible that work in the area to the east may show that this fault is an extension of the one described in the south end of the Black Pine Range. If so, its course is very sinuous and must exceed 50 miles in length.

EARLY TRANSVERSE AND NORMAL FAULTS

Following soon after, or perhaps in part overlapping, are transverse faults and normal faults which belong to the same orogenic epoch in which the folding and thrusting occurred. These have dislocated or offset the earlier structures, but are themselves affected by a still younger epoch of faulting, probably to be correlated with the general Basin-Range faulting. Only the most important of the earlier faults will be described.

KELSAW FAULT

The largest and most important of the transverse faults is that which separates the strata of the middle section of the Black Pine Range from those on the south. This fault has pronounced topographic expression and the carving of Kelsaw Canyon has been mainly directed by this fault, and also the relatively low divide between it and the head of Black Pine Canyon (Plate IV, A, and Plate X, A). The trace of the fault is not now directly in the floor of the canyon, but lies a short distance up the north slope. Its general direction is due east, but topographic inequalities cause an apparent swing high up the middle section of the range, around the north side of War Eagle Peak. Descent then occurs into East Dry Fork Canyon and thence over the divide into Mineral Gulch and eastward to the base of the mountain.

All folds and all other faults end directly against this fault. Its general character is normal, with downthrow on the south, which has brought the overthrust block of the Brazer formation alongside strata of the middle and upper Wells. It is very unlikely that the fault has a large horizontal component of movement, for structures on opposite sides of the fault plane do not match. The amount of movement in a horizontal direction cannot be estimated, but the vertical component is probably two or three thousand feet. The dip of the fault plane is about 35° to the south. Beds on the south side of the fault plane have a general inclination down and against the fault plane, as shown in structure section O'-O''.

OTHER FAULTS

Normal faults in the middle section of the Black Pine Range were discussed in describing the folding. The largest ends on the south against the Kelsaw fault, but extends northward for two and one-half miles before disappearing beneath the alluvium at the edge of the mountain. The curious pattern of Pole Canyon is largely due to this fault. The upper mile of the canyon is directly on the fault, but as the canyon turns from northwest by north to directly east, the fault continues its northerly direction, as indicated by the low saddle across the divide into Sweetzer Canyon (Plate IV, B). It then turns slightly to the east and is in the main closely followed by Sweetzer Canyon to the edge of the alluvium. Downthrow is to the east and the dip of the fault plane is probably more than 45°, as suggested from the exhumed steep triangular facets across the ridges, as illustrated in Plate IV, B. Displacement is probably between 500 and 1,000 feet. The fault is essentially a strike fault. A fault across the ridge in Mill Canyon a half mile to the east is similar in every way. However, Mill Fork does not remain on the fault, but swings to the east, while the fault, although crossing from East Dry Canyon over a low saddle into the head of Mill Fork, is outlined by a series of saddles across tributary ridges on the eastward-facing slope of Mill Fork Canyon. The fault crosses Pole Canyon, outlined by a depression or saddle on the ridge to the north, and passes beneath Tertiary and Quaternary deposits at the mouth of Jones Canyon. It also has a downthrow on the east (structure sections L-L' and M-M').

Faults in the Sublett Range have also been mentioned. The one followed more or less closely by Sublett Creek is possibly a transverse fault of considerable magnitude (structure section K-K'). Downthrow is apparently on the north. Vertical displacement is probably about 1,000 feet, but the horizontal component may be much greater. Further study east of the district is necessary for final interpretation. The fault followed by Cold Spring Canyon is parallel to the one just mentioned. Its vertical displacement is about 500 feet and the downthrow side is on the north. Other similar faults doubtless exist in the Sublett Range. Possibility of a strike fault extending from near the reservoir at the mouth of Lake Fork and passing to the northwest beneath Tertiary strata has been cited elsewhere, and it may be necessary to explain the apparent duplication of the Phosphoria formation.

A transverse fault of considerable magnitude may exist in the Albion Range along the southern base of the northern section. Its relations, however, are not sufficiently well known to make a positive assertion. In fact, several faults may exist and the topographic low across the range may in part be the result of more effective erosion in a much-faulted zone. Such faulting may also have been most effective in bringing the underlying Paleozoic block to view and thereby exposing the overthrust character of the Albion Range.

YOUNGER NORMAL FAULTS

The distinction of younger normal faults is made because those already described are older than the Tertiary strata which have not been broken by them, whereas those to be described have displaced the Tertiary sediments and lavas. Most of these faults are outlined by pronounced scarps, but others are mainly inferred from stratigraphic relations. These faults occur on both sides of the Albion Range and have caused the tilted block ranges in the district.

MALTA RANGE FAULT

Most impressive of the younger faults is that which has blocked out the Malta Range. Many of its features have already been described or discussed in the section on topography. Mention was there made of the appearance of the range as a tilted fault block with steep eastward-facing scarp, as shown in Plate XIV, B, rising gradually from the ends of the range to an elevation of 3,000 feet above the alluvial-floored Raft River Valley near its mid-point. Mention was also made of the gentle west slope of the range. This fault, which borders the east side of the range, trends about due north and apparently dips steeply east (structure sections A-A', B-B', D-D'). Its magnitude remains undetermined, and, although 3,000 feet is shown in the scarp near the center of the range, the displacement is probably much greater. Judging from the westward tilt of the lava on the east side of Raft River Valley, the downthrow side must lie below several thousand feet of alluvium. Steepness of the original fault plane has resulted in numerous landslides, some of the blocks of large size. As a result there has been a decided retreat of the scarp in the higher parts of the range, but in parts of lesser elevation the scarp has been but little dissected and the range as a whole is characterized by its extreme youthfulness. The fault scarp descends gradually northward from its highest point until finally the entire range plunges beneath the Snake River Plains and the fault plane is concealed by the young Snake River basalts. That its trace is continued beyond the end of the range below the floor of the lava-covered plain is suggested from the alignment of a basalt vent several miles north of the range and on a projected continuation of the fault trace. Several small basalt vents also occur close to the base of the scarp near the north end of the range. So far as known, one hot spring ascends along the fault plane near the south end of the range.

Another fault has also played a role in the development of the Malta Range. This lies at the west base of the range, but unlike that on the east side has no topographic expression in the range itself. Its presence is mainly concealed by alluvium at the base of the range, but scarps appear in the extreme north and south ends on the west side of the fault plane. The fault is apparently similar to that on the east side of the range, with the Malta Range the downthrow side and the Albion Range the upthrow side. A branch of the fault plane is more or less closely followed by Marsh Creek northeast of the town of Albion, where a stratigraphic throw of about 1,500 feet is represented. This measurement is from the lava surface at the base of the Malta Range to the lava surface that caps the north end of the Albion Range. Its scarp has much the same appearance as that on the east side of the Malta Range. Southward the Tertiary rocks have been stripped from the upthrow side, leaving only pre-Cambrian rocks. Lava on the west side is again exposed west of Elba where it caps the Carboniferous and pre-Cambrian rocks about 1,500 feet above the base of the tilted block east of Elba. Erosion has here also removed the Tertiary rocks from the flanks of the Albion Range between Elba and Almo, but a scarp with the higher side of tuff capped by lava reappears four miles southeast of Almo in alignment with the fault indicated at the west base of the Malta Range. Additional data confirming its presence are afforded from remnants of latite on the high ridge west of the Cassia City of Rocks area at nearly 7,000 feet A.T., and also by a small remnant in the south end of the rock city itself. Displacement along the fault plane is apparently not so great as along that on the east side of the Malta Range.

OAKLEY FAULT

Another fault of much lesser magnitude but with a pronounced scarp (Plate XV, B, structure section C-C') appears several miles in front of the west base of the Albion Range or about two miles east of Oakley. This fault also cuts the lavas and is similar to the Malta Range fault, except that its relations are reversed; that is, the scarp faces westward and the back slope is to the east. The tilted block rises about 500 feet above the alluvium-covered floor of Goose Creek Valley and may be traced for 10 miles in a generally northerly direction, beginning near the north end of South Mountain. Only flows of latite appear in the scarp, and for that reason landslides have not seriously marred its face. The fault plane dips steeply west and the back slope about 4° to the east. The total displacement cannot be estimated, but is probably a thousand feet or more. In the same region are several other similar faults of apparent lesser displacement.

SOUTH MOUNTAIN FAULT

South Mountain has also all the characteristics of a tilted fault block or in actuality a horst block (structure sections D-D', E-E' and F-F'). Its eastern slope rises abruptly from the wide basin of Birch Creek and Junction Creek and it has a gentle slope to the west. As discussed in the section on topography, the drainage divide as well as the axial line of the range is at the top of the steep eastward-facing scarp. Additional evidence of faulting is also afforded by stratigraphic relations, for the strata at the crest of the range or at the top of the scarp are the same as those at the west base of the Albion Range directly across the valley. In both places the quartzites form a thin shell above intrusive granite, and, inasmuch as the faulting is later than the intrusion, the granitic rock is widely exposed in the steep fault scarp of South Mountain beneath the thin shell of metamorphic rock. Should South Mountain be depressed about 1,500 feet the rocks would almost exactly match the west slope of the Albion Range. Displacement along the fault is therefore about 1,500 feet with downthrow on the east. The dip of the fault plane probably exceeds 60°, as much of the scarp is very precipitous. The other side of the block slopes westward at angles from 10° to 12°. The fault plane is curved slightly to the west from true north. Movement along it may have been initiated before deposition of the Tertiary formations, but considerable displacement has taken place along it subsequently.

Another fault also bounds the range at its west base, this fault with downthrow to the west. Both faults end against a northwesterly trending fault at the north end of the range, one which might be regarded as a cross or transverse fault. This fault has downthrow to the northeast, and like the others, is of normal character. The three, acting about contemporaneously, caused the South Mountain block to develop as a horst and to become detached and isolated from the main Albion Range. At the same time several other northwesterly trending transverse faults were developed which caused slight topographic offsets of the range along the course. The one now followed by Cottonwood Creek and Cold Creek has caused a very distinct interruption or offset of the range, as well as of its structural trend. These features are well portrayed on the geologic map.

OTHER FAULTS

A normal fault of great extent and magnitude probably lies at the west base of the Albion Range, but evidence of its presence is not so obvious as in those faults already discussed. The fault is believed to extend from near the north end of South Mountain to the end of the range east of Declo. Evidence for its presence is inferred from both topographic and stratigraphic relations. Faulting along the west base of the Albion Range is at once suggested by the great regularity of slope and the lack of indentation such as normally characterizes erosional slopes. As seen from the west the range appears as a bold, high, ridge-like scarp, springing abruptly from the broad alluvial plain at its base and holding a line of simple curvature. Short canyons on the steep slope hold their sharp V section to the mountain base. This slope and simple curvature of the base line is similar to that along the west front of the Wasatch Range in Utah. Erosion has carved no farther into the windows of more readily erodable Paleozoic rocks at the margin of the range than in the much more resistant metamorphic rocks. East of Oakley and a few miles to the north, the latites dip against the range. It is probable, as the scarp line suggests, that the Tertiary strata on the west have been faulted against the older rocks on the east. Latite and tuff on the range northwest of Albion are cut off abruptly on the west side of the range. Alluvium obscures all relations at the base of the range 500 feet below. Latite on the range west of Elba is nearly 2,000 feet above similar latite at the west base of the range east of Oakley. These stratigraphic interruptions give rather conclusive evidence of the faulting which the topographic features so strongly suggest. A basalt vent or dome south of Burley is along a nearly parallel fault of similar character and its presence further supports the inference of the main fault and its probable normal character. The course of the fault has simple curvature to the northeast. It disappears on the north as the Albion Range plunges beneath the Snake River Plains and also ends near the north end of South Mountain. Its development was contemporaneous with the faults which blocked out the South Mountain horst. Near the southern end it loses its simple plane curvature as a consequence of the contemporaneous development of the South Mountain block, and the movement is taken up by several faults having a northwesterly direction, as shown on the geologic map. These are cut by subsidiary faults parallel to the one that bounds the north end of South Mountain. This somewhat peculiar pattern into which the main fault passes may be ascribed to shearing forces developed as the South Mountain horst moved relatively upward on the one hand, with downthrow on the north, while the depressed block at the west base of the Albion Range was being lowered in contra-position. As a consequence of the differential movement, or differential application of forces, the rocks of the immediate and nearby zone were broken by faults directed along planes of maximum shear or at 45° to the course of the fault farther north. Bending of the Oakley fault plane as it enters the shear zone is strikingly shown on the geologic map.

Displacement along the fault probably is not everywhere the same, but in places attains 2,000 feet and probably in others twice as much. Dip to the northwest is probably near 60°.

There is much faulting of less magnitude in the Albion Range. Some of the faults cross the range, others are directed along it. One bounds the west side of the Albion Basin. It probably continues across the high part of the range to the south and it may have faulted the old erosion surface on the summit of the range, thereby causing the appearance of the second high level on the top of Mount Harrison.

Several other faults of much smaller magnitude are also present in the district, but these warrant only passing notice. One is near Churchill about eight miles north of Oakley, where a small block of latite rises less than 100 feet above the surrounding alluvial plain. It has a steep slope to the southwest and a more gentle one on the northeast. Unlike the other young faults in the district, this one strikes northwest, but has downthrow on the southwest side. This fault may be traced for nearly two miles, with the ends concealed by alluvium. A similar fault was seen near the east edge of Raft River Valley about three miles southwest of Heglar, but the strike was more nearly east and west. A block of latite there rises about 150 feet above the surrounding formations and has a scarp facing the south and a gentle slope to the north. Normal faulting belonging to this epoch has affected the strata east of the Raft River Valley but very little, nor has faulting been prominent in the region west of the district.

LANDSLIDE FAULTS

Minor faults unrelated to major tectonic movements are abundant in the district, and owe their presence to the force of gravity. Such faults are especially abundant in the face of the steep normal faults which cut the Tertiary strata and are particularly numerous in the front of the Malta Range. These faults are the planes of movement over which the landslide blocks have broken loose from the steep major fault scarps and have slid toward the base of the range. The tuff beneath the thick heavy lava capping is especially favorable to such land movement. These faults are not confined to the vicinity of the large normal faults, but occur also in gulches or canyons which have cut through the lavas into the underlying tuff. Landslide faults also occur elsewhere, especially in the over-steepened valleys of the Blackfoot cycle in the Sublett Range. These faults are shallow surface features.

SNAKE RIVER DOWNWARP

Most interesting, as well as the most peculiar, of the structural features of the district, is the Snake River downwarp to which frequent reference has been made. It represents a great structural depression or downwarp which stretches entirely across the State from the Wyoming to the Oregon line, superimposed on all previous structures. The Tertiary strata and older rocks, including the mountain ranges composed of them, are tilted gently toward the center of this great downwarped area. The essential features have recently been summarized by Kirkham.[1] Upon this downwarp or geosyncline have accumulated younger sediments, but more impressively, the basalt flows of the Snake River Plains.

In Cassia County reference has already been made to the way the north ends of the Sublett, Malta, and Albion ranges disappear by plunging beneath the constructional Snake River Plains. Erosion surfaces as well as the structural features are carried below by the warping. In turn, complementary movement has arched the ranges farther south to higher elevations and has produced a broad anticlinal structure parallel to the downwarp. Flanks of the downwarp dip from 5° to 8° to the north to the margin of the Snake River Plains.

AGE OF THE FOLDING

Strata involved in the early folding range in age from pre-Cambrian to Permian. These folds are essentially continuous with similarly folded structures in southeastern Idaho, where Mansfield[2] has found beds as young as the Wayan formation of Lower (?) Cretaceous age folded to

---

1 Kirkham, V.R.D., Pamphlet 19, op. cit., pp. 24-26; also Jour. Geology, Vol. 39, op. cit., in press.
2 Prof. Paper 152, op. cit., pp. 169-172.

the same degree as the older formations. He concludes that the main epoch of deformation which produced these folds was later than the deposition of the Wayan Beds and probably a part of the great mountain-building disturbance which came in the interval between the end of the Cretaceous period and the early part of the Tertiary, generally known as the Laramide revolution. Some evidence of other earlier epochs of mountain-building is furnished by the rocks in that region, but the effects of these earlier disturbances were largely obliterated by the more intense activity of the Laramide revolution. Similarity of folded structures in Cassia County suggests folding during the same epoch. That this is actually so was verified by tracing the folded structures more or less continuously into the area described by Mansfield.

A subsequent epoch of less extensive disturbance is suggested by the arching and tilting of the planes of the overthrusts and by the dips now observed in the Miocene strata and associated lavas. On the present supposition that the Salt Lake formation is of Pliocene age, Mansfield assigned the disturbance to the end of the Pliocene. Although the Tertiary rocks may be older than the tentative age assigned by Mansfield, or Upper Miocene instead of Pliocene, there is no reason for assigning any other date to this disturbance than the one given. Further discussion of Pliocene and more recent diastrophism will follow in the section on the age of the faulting.

The erosional history indicates that there were several disturbances in the Tertiary and Quaternary, each in the nature of broad regional doming or uplift. Apparently folding was involved only in the movement that came near the end of the Pliocene. Folding at the time of the Snake River downwarp will be discussed later.

AGE OF THE FAULTING

Mansfield has also ascribed the Bannock overthrust and other thrust faults in southeast Idaho to the Laramide epoch and to the later part of that epoch, because the faulting did not occur until the rocks in what is now the upper block had been intensely folded—indeed, folded practically as much as those of the lower block. The Albion, Sublett, and Black Pine overthrusts are similar in these particulars, for they, too, came after the strata in both blocks had been intricately folded. From the striking similarity of the faulting in the two regions, it seems almost certain that all were produced during the same epoch. Those in Cassia County probably also occurred late in the Laramide disturbance. The time when the fault planes were folded is uncertain, but it is believed that folding was much later than the production of the thrust faults themselves.   Possibly it was a part of the late Pliocene disturbance.

Many of the normal faults and also the transverse faults cut the overthrusts and are therefore younger, but these same faults have been buried by Miocene strata and are therefore older than the mid-Tertiary. They should probably be given a place late in the Laramide epoch or immediately following it.   These faults may be the result of a tensional phase, which probably followed the intense compression experienced by the region in the Laramide deformational epoch, during which the deformed mountain mass slowly returned to equilibrium by means of the normal or relaxational faults.

In nearby regions normal faults are described as belonging to the epoch of block faulting which gave rise to much of the so-called "basin and range" structure, of which Utah and Nevada furnish the best examples.   This structure is generally stated to have been developed in about Middle Miocene time.   There is good reason for believing, however, that most of the faulting is much younger and is late Pliocene or early Pleistocene in age.   It is possible that some faulting of this character took place in the Miocene and inaugurated the disturbance which intiated the Upper Miocene sedimentation, but if so, trace of such faults has subsequently been lost beneath the Tertiary capping or obscured by younger movements along the same fault planes.

Normal faulting which has caused the present block ranges in Cassia County and which has outlined the Albion Range in its present shape, is younger than the Miocene strata and the overlying latite series. These faults are comparatively young, as the fault scarps are fresh and little eroded except where high elevation and the exposure of the tuff series has specially favored landslide action.   Such faulting preceded the glaciation in the Pleistocene, as the fault block ranges themselves show effects of high mountain glaciers, but the faulting followed after considerable erosion of the Miocene strata and lavas.   It is certain that these faults were generally later than the Gannett erosion surface, which was developed late in the Pliocene,[1] for faulting has destroyed the erosion surface as well as one or two of the earliest Pleistocene surfaces.   In fact, none of the fault block ranges bear the remnants of the erosion surfaces, perhaps because the active erosion following the relatively sudden rejuvenation was not favorable to their preservation. Faulting may have initiated or been in part responsible for renewal of the erosional cycles.   Faulting was largely completed, however, before the inauguration of the Blackfoot cycle, for this cycle has been active

[1] Mansfield, G. R., Geography, geology, and mineral resources of the Portneuf quadrangle, Idaho: U.S. Geol. Survey Bull. 803, 1929, pp. 7, 66-67.

in dissecting the fault block ranges. Although some faulting may have occurred late in the Pliocene, it is believed that most of the movement or displacement occurred rather early in the Pleistocene and that the fault block ranges are essentially early Pleistocene in age. That some movement has also continued into later times is suggested from the presence of a small scarp in the Snake River basalt flow several miles east of the north end of the Malta Range.

AGE OF THE SNAKE RIVER DOWNWARP

Kirkham was first to show that the warping in the eastern Snake River Plains area was later than the "Tertiary Late Lavas" of supposed Pliocene age.[1] In more recent studies in southwestern Idaho, he has obtained additional data and has found that not only is the downwarping subsequent to the deposition of the Upper Miocene Payette beds, but mainly subsequent to the deposition of much of the Pliocene Idaho formation.[2] Subsidence thus began in the Pliocene, was active in the Pleistocene, and has continued into recent times.

The evidence in Cassia County also supports the late age for the downwarp, probably mainly in the early Pleistocene. Although subsidence may have begun earlier in the more central parts of the basin, the present marginal slopes involve not only the late Pliocene or early Pleistocene tilted block mountains, but also the Gannett, Elk Valley, and probably the Dry Fork erosion surfaces, the last two of Pleistocene age. Subsidence had largely ceased, however, before the advent of the glacial stages and more particularly before the completion of the Blackfoot erosional cycle.

CAUSE OF DIASTROPHISM

LARAMIDE DEFORMATION

The intensity of folding and faulting during the early epoch of deformation can be ascribed only to great tangential pressure. No better cause for this deformation has been found than the segmental hypothesis of T. C. Chamberlin[3] and the closely related wedge theory of diastrophism by R. T. Chamberlin.[4]

On the basis of field measurements in the Appalachian Mountains of Pennsylvania, R. T. Chamberlin has propounded a wedge theory for the formation of mountain ranges and of continents. The continental

---

1 Pamphlet 19, op. cit., p. 26.
2 Kirkham, V. R. D., Snake River Downwarp, Jour. of Geology, Vol. 39, 1931, pp. 471-482.
3 Chamberlain, T. C., and Salisbury, R. D., Geology, Vol. 1, 2d ed., pp. 542-549, 1905.
4 Chamberlin, R. T., The wedge theory of diastrophism: Jour. Geol., Vol. 33, pp. 755-792, 1925.

and suboceanic masses are treated as segments, the suboceanic masses being the larger and heavier. The principal source of the deforming force is the rearrangement of material in the interior of the globe in favor of greater compactness and higher density. This should cause general shrinkage, with the resulting circumferential compressive stresses beneath both continental and oceanic areas. In the shrinking process all segments would sink, but the master segments would take the lead and would squeeze the smaller and lighter continental segments between them so that they would be wedged upward. As applied to mountain chains it tends to produce more or less symmetrical, marginal folded areas, bounded by exterior, inward dipping shear zones and a relatively undeformed intermediate area. For those mountains in which the greatest effects of overthrusting appear on the inland side of the range, R. T. Chamberlin believes that the mountain-built area was caught between an oceanic segment on one side and a continental segment on the other, and that the pressure of the oceanic segment tended to overbalance that of the continental. Localization of the deformation would be in the deeply loaded geosynclines at the margins of the oceanic and continental segments, where weakness of the earth's crust would be the greatest.

The entire region here described is part of a great geosyncline in which sediments were deposited with few interruptions of importance from Proterozoic to late Mesozoic times. This great structural feature has been called by different writers the Rocky Mountain trough, the Laramide trough, or the respective geosynclines. It extended from the Arctic Ocean southward through the Great Basin, and was in general an area of subsidence in which sediments had accumulated in great thickness. On the west, throughout much the same interval, a relatively persistent land mass had separated the geosyncline from the Pacific Ocean, and on the east a less persistent barrier at times had separated it from the interior seas. Ultimately tangential pressures acting from the west-southwest overcame the resistance of the greatly weakened trough of sedimentation, compressed the strata into folds and caused great overthrusts. Shortening of the crust by folding and thrust faulting must have been considerable, but no estimate of the amount can be attempted here. The frequency of inclined or overturned structures and the original nearly horizontal attitude of the overthrusts indicate that the forces acted horizontally and were not the surface expression of deep-seated shear. Because of the great amount of crustal shortening, as evidenced by the folding and thrusting, the mountains produced must have been of the "thin-shelled" type, distinguished by R. T. Chamberlin[1] as those in which the intensely compressed outer portion of the earth's shell has sheared upon a less yielding base beneath, without disturbing the earth's crust to any great depths. Great overthrusts are marginal features of the mountain-forming area of the thin-shelled type of deformation. The folds and overthrusts in Cassia County and to the east represent therefore the east margin of a wedge zone as distinguished by Chamberlin, and of a wedge asymmetrical with respect to vertical lines. Where the opposite side of the wedge makes surface expression is not certainly known. Such thin-shelled mountain zones are, according to R. T. Chamberlin,[2] accompanied by little igneous activity of any sort in the marginal and most strongly overthrust portions; but in the heart of the deformed belts, where there has been more uplifting and the affected zone goes deeper, granitic and other intrusions are a common and probably characteristic feature. With this point in mind, one might suggest that Cassia County is well toward the inner part of the wedge, inasmuch as the granitic intrusions have here made their appearance.

The intense compression suffered by the region during the Laramide deformation was doubtless succeeded by a condition of tension, during which the overstrained mountain mass slowly returned to equilibrium. Normal faults in regions that have undergone folding commonly mark a late stage of that deformation. Apparently the folded rocks have in large measure been strained beyond the limits of their stability, and equilibrium has resulted when the rocks tended to resume their former positions, and thus create a strain, which, in turn, is relieved by normal faulting. Such are the current views regarding the origin of normal faults in relation to the folded chains.

BASIN-RANGE FAULTING

Many hypotheses have been offered in explanation of the Basin-Range structure, either involving tensional forces or compressional forces. Some explain the fault blocks as upthrusts due to deeply applied compressive forces, some to the collapse of a gently folded or arched dome. In Cassia County, tensional forces seem best to account for the observed facts, for the evidence such as simple curvature of the fault traces, the occurrence of hot springs near the fault plane, and the near occurrence of basalt vents, all favor normal faulting and not thrusting. which would tend to close fractures and make them impermeable. Faulting of normal character or of the Basin-Range type possibly

---

[1] Chamberlin, R. T., The building of the Colorado Rockies: Jour. Geol., Vol. 27, pp. 248-251, 1919.
[2] Chamberlin, R. T., Volcanism and mountain-making; Jour. Geol., Vol. 29, pp. 166-172, 1921.

began or occurred in the mid-Tertiary as in other parts of the Great Basin country, but activity was revived late in the Tertiary or more probably early in the Quaternary with the development of the present striking block ranges in Cassia County.

These later deformative epochs were probably begun by revival of periodic contractional disturbance of the earth. But at these times the effects were not so definitely localized as before, probably because no accumulation of sediments comparable to those of the former geosyncline had taken place, and the rigid crust was therefore not weighed down as before. The result was a broad uplift with only gentle foldings and warpings. The broad uplift and renewed folding of the later deformational epochs were also succeeded by tensional phases, apparently more marked than those that followed the Laramide epoch. R. T. Chamberlin[1] also distinguishes "thick-shelled" mountains which show open folding, but are without thrust faults or evidences of intense horizontal compression. In these mountains the depths affected by the compressive forces are far greater than those of the "thin-shelled" type, and as additional contrast the "thick-shelled" mountains seem to be more commonly associated with volcanism than mountain-making movements which affect a thin shell. Outpourings of lava are especially common. The younger epochs of deformation in Cassia County and the Great Basin country in general seem to be the "thick-shelled" kind and have been accompanied by vast floods of lava. Extrusions of latite and some basalt may be considered as associated with the tensional phases of the mid-Tertiary doming and partial collapse by normal faulting, and the later outpouring of basalt with the tensional phase of the late Tertiary or early Quaternary doming.

The Basin-Range structure is therefore the probable result of normal faulting and partial collapse of a region following broad uplift or arching, and the fault blocks represent the readjustments made in the overstrained shell in its attempt to recover equilibrium. The normal or block faulting also afforded opportunity for the outflow of lava, because the thick-shelled movements probably extended to great depths, as suggested by R. T. Chamberlin, and have very likely reached the zone of potential liquefaction of rocks, a depth seldom attained by deformation of the "thin-shelled" kind. Or it is possible that magmatic action is mainly responsible for the faulting and the block ranges are the result of jostled blocks over a magma reservoir. The nature of the fault pattern, especially along the west margin of the Albion Range, is highly suggestive of such jostling, with some blocks settling, others

---

[1] Jour. Geol., Vol. 27, pp. 248-251, 1919.

rising, with development of additional shearing strains in zones between rising and subsiding blocks. The great normal faults, whose directions of trend are to the north, are obviously a part of the Basin-Range tectonics. Location of the basalt vents along these fault planes suggests a strong genetic relation between the faulting and the movement of magma, and that the jostling is due to the readjustments to movements of magma from one part of the reservoir to another, or to the surface.

SNAKE RIVER DOWNWARP

Causes given to account for the Laramide and Basin-Range deformations fail to explain the Snake River downwarp. This feature is a negative element, a subsiding trough or geosyncline in which sediments and lavas have accumulated and are still accumulating. Compressional forces fail to explain the downwarp. Its alignment is not parallel to the direction of forces generally associated with a shrinking earth and the depression of oceanic and continental segments. The general lack of faults about its margin, the simpleness of its curvature can give no other interpretation than subsidence or downwarp. Such subsidence is not due to the accumulation of load, for the load has come later. The suggestion made by Kirkham[1] that the downwarp is the result of withdrawal or migration of magma perhaps affords the most suitable explanation. Fractures reopened during the warping have also permitted floods of basaltic lava, the Snake River basalt, to come to the surface and partly fill the basin.

GEOLOGIC HISTORY

PROTEROZOIC ERA

There is nothing which might give a clue to the history of the region prior to the Proterozoic era, but at some time, probably rather early in the Proterozoic, there was prolonged marine sedimentation, during which the members of the Harrison series were deposited.

Much of the material of the Harrison series consists of clean quartzites, derived mainly from fine sands, along with an occasional thin pebble bed. The general uniform assortment of these sands together with their purely siliceous character and constancy of well defined uniform bedding, suggests strongly the effect of long-continued wear of the materials, probably such as takes place on a beach or in shallow water constantly agitated by the waves. Such deposition could only have

1 Kirkham, V. R. D., Snake River Downwarp, Jour. Geol., Vol. 39, 1931, pp. 456–482.

taken place on a slowly sinking sea floor in which subsidence kept nearly exact pace with deposition, for during this time a thickness of more than a mile of these shallow water sands accumulated. Several times accumulation of sand was interrupted by an increase in the rate of subsidence, and each time only the finer materials could reach the region of deposition below the zone of wave action. Each time sediments more or less muddy were laid down, which subsequently have been changed to the micaceous quartzites and the thin beds of mica schist now found in several places in the lower division of the series. Interruption of sand accumulation was more frequent later in the period, for the middle division is characterized by an accumulation of argillaceous and calcareous beds. Deposition of the muds and silts could have come only with the deepening of the water, and the limestone only in the deeper water sufficiently offshore so that it could not have received notable admixture of clastic materials. The region of limestone deposition must have been near enough land at times, however, to receive considerable quantities of land-derived sediments, and shoaling of the sea must have occurred several times to account for the alteration of sand, shale and limestone. Near-shore or shallow water conditions must have again prevailed near the close of the epoch of sedimentation, for the last sediments are much the same as those that accumulated earlier in the lower part of the series. It is not known just when sedimentation ceased and land conditions prevailed, for the upper part of the series has been eroded. Suffice it to say there was an accumulation of marine shallow water deposits nearly two miles thick and that the part now exposed is probably far from the complete section.

It is likely that at the cessation of sedimentation the newly deposited strata were subjected to orogenic movement and the rocks considerably metamorphosed and perhaps changed into much their present state. Such an assumption is probably justified in view of the slight metamorphism shown by the rocks assigned to the upper part of the era in nearby areas. Apparently the strata were not greatly folded at the time, for even today the series has been but simply arched and because of its competent character less folded than the younger Paleozoic formations.

Sedimentation was probably resumed at a later date, but inferences regarding the upper Proterozoic era must be drawn from the evidence furnished by neighboring regions. It is possible that the district was land during most of the upper Proterozoic or Beltian epoch, but it is more likely that there was continental deposition such as occurred in the Wasatch Mountains and in the northern part of Idaho and in western Montana. No doubt the Rocky Mountain or Cordilleran geosyncline was well advanced during this period, indeed may have had its beginning long before the Harrison epoch. According to Walcott,[1] this geosyncline was well advanced during the Algonkian period and then extended from the head of the Gulf of California northward probably to the Arctic Ocean, and was the seat of prolonged sedimentation which he regarded as non-marine. This favors continental deposition in the region of Cassia County, and as elsewhere, deformation, erosion, and even perhaps glaciation. Glaciation is suggested, for in the Big and Little Cottonwood Canyons in Utah, Blackwelder[2] describes tillite as an important and interesting member of the upper pre-Cambrian series. Deformation is also suggested, not alone from the difference in the degree of metamorphism between the upper pre-Cambrian and the overlying Paleozoics in the same region, but also from an angular unconformity that Blackwelder finds between the Algonkian and the overlying Cambrian quartzites in Cottonwood Canyon south of Salt Lake City. In the later part of the Proterozoic and in early Cambrian time the geosyncline must have thus been subjected to erosion, which reduced it nearly to base level, and there was apparently little warping needed in Cambrian time to permit the flooding of the geosyncline by marine waters.

PALEOZOIC ERA

In the later part of the Lower Cambrian or in the early part of the Middle Cambrian progressive subsidence of the Rocky Mountain or Cordilleran geosyncline was again resumed and the trough began to receive sandy sediments, at first probably non-marine but later marine. Slow subsidence continued until probably several thousand feet of shallow-water sands had accumulated, derived no doubt from erosion of pre-Cambrian quartzites farther to the west. Later the waters deepened and the deposition of clastics gave way to the deposition of an even greater thickness of limestones lasting through much of the Middle Cambrian and most of the Upper Cambrian.

There is no record of subsequent events within the district until well up in the Carboniferous. In the interval the series of events is probably much the same as recorded elsewhere in the general region, where the record is dominantly marine and indicative of progressive subsidence of the geosyncline. The variation in the character of the sediments, however, and the numerous stratigraphic breaks show that

[1] Walcott, C. D., Cambrian geology and paleontology, III: Smithsonian Misc. Coll., Vol. 64, pp. 50-81, 1914.
[2] Geol. Soc. Amer. Bull., Vol. 21, pp. 517-542.

there were at times interruptions in subsidence and even reversals of movement, with erosion. These disturbances were gentle, but were fairly well distributed throughout the interval.

According to Mansfield,[1] withdrawal of the sea from the geosyncline probably occurred at the close of the Cambrian, followed by brief erosion, but limestone making again took place early in the Lower Ordovician. Later a reversal of subsidence in the later part of the Lower Ordovician time brought first a change in the sedimentation, the deposition of sands, and then an erosional interval that lasted throughout much of Ordovician time. Toward the end of the Ordovician subsidence was resumed and limestone deposition began again. Cessation of sedimentation probably occurred at the end of Ordovician time, but deposition of limestone was again resumed in the Silurian after a stratigraphic break of some magnitude. Toward the end of the Silurian the sea withdrew and did not again reflood the geosyncline until Middle Devonian time, when limestone making was resumed without any marked discordance in the attitude of the later with the earlier beds. Limestone deposition continued to the close of the Devonian, when there was again emergence followed by slight erosion, but subsidence again followed and limestone making was resumed early in the Carboniferous.

The lowest of the Carboniferous strata is not exposed in Cassia County, but from the record farther east limestone was probably deposited during the Lower Mississippian (Madison limestone). Withdrawal of the sea for a time at the close of the Lower Mississippian is suggested from the lack of exact conformable relations with the overlying Upper Mississippian strata, the Brazer formation. In the Upper Mississippian the record in Cassia County again becomes available. Throughout much of this time clear seas prevailed, and in them much limestone accumulated. Several times, however, the limestone deposition was interrupted by reversal of subsidence, and the sea became muddy and muds or silts were deposited. At times the sea became shallow and wave action so agitated the water that only sands could accumulate. The site of deposition was probably much nearer the shore of the Upper Mississippian sea than in the region farther east, where the entire accumulation consists dominantly of limestone. It thus received more notable contributions of land-derived clastics. Perhaps the changes in character, or alteration in characters of the formation, might be ascribed to oscillations of the neighboring land mass and repeated rejuvenation with resulting increase in the quantity of clastics

[1] Prof. Paper 152, op. cit., pp. 173-188.

carried to the sea. According to Mansfield the sea again withdrew from the geosyncline at the close of the Mississippian, but little disturbance of the land took place, for with the return of the sea in Pennsylvanian time the deposition of limestone was resumed under much the same conditions as in the preceding period and without marked discordance in bedding. Pennsylvanian time records, however, much greater oscillatory movement of the sea bottom than previously. Limestone deposition prevailed at the beginning and at the close of the period, but even during the lower part and especially during the middle of the epoch shallow waters prevailed and favored the deposition of sand. Periodic rejuvenation of the bordering land mass may also have favored the accumulation of detrital material, but this seems less likely than movement of the sea floor because of the lack of intercalated shale members, but instead there is the direct passage of limestone to sandy limestone and to sandstone. For a considerable time in the middle of the epoch little except sand was deposited. The sea withdrew at the close of the Pennsylvanian, but when the sea returned in the succeeding Permian period conditions of deposition had greatly changed. In the first place inundation probably came from the north or west instead of from the southwest as in previous floods, and under special conditions of deposition not yet well understood, phosphatic shales and thick beds of chert were deposited. Withdrawal of the sea is presumed to have occurred again at the close of the Permian.

MESOZOIC ERA

The Mesozoic record in Cassia County is very imperfectly known. No strata belonging to this era have as yet been identified and it is possible that throughout much of the time the region constituted a land mass undergoing erosion. The latter part of the era, however, witnessed great diastrophic movements and these have left a notable record in the district.

Mansfield describes prolonged sedimentation during the Mesozoic era in the region to the east and this sedimentation, no doubt, has had some reflection in Cassia County. As a whole the Mesozoic is there characterized by alternating marine and non-marine deposition with the latter dominating toward the end of the era. Marine embayment continued throughout Lower Triassic times and was followed by an epoch of erosion and the deposition of continental red beds, chiefly of the desert type. The change from the Paleozoic to the Mesozoic was marked by faunal rather than by any noteworthy discordance of strata, and for this reason one might suppose that the Lower Triassic seas also extended over Cassia County. Continental deposition continued through Lower and Middle Jurassic, but in the Upper Jurassic marine invasions twice alternated with desert continental conditions. An abrupt change in the character of the sediments marked the transition to the Cretaceous, and, according to Mansfield, such as to suggest mountain building in neighboring areas. As Cassia County is much nearer the site of the late Jurassic Nevadian revolution than the region to the east, it no doubt was very near the border zone of the disturbance, if not within. The Lower Cretaceous witnessed the accumulation of an enormously great thickness of fluviatile, together with some lacustrine, beds. It is more than likely that Cassia County was a land mass after a withdrawal of the sea in the Lower Triassic. It may have been at times the site of continental deposition, but for most of the time it probably contributed to the great thickness of continental deposits to the east, especially toward the close of the era. In southeastern Idaho the Upper Cretaceous was a time of erosion.

The Laramide revolution closed the Cretaceous period with the deformation of the geosyncline and the development of the Rocky Mountains, with its complicated structural features. Seemingly the accumulation of tangential crustal pressures finally overcame the resistance of the heavily loaded geosyncline and these pressures, acting most strongly from the west and southwest, compressed the great thickness of strata into a series of northwestward trending folds, generally overturning them, and then slicing them along great low-angled fault planes. Possibly, when the thrusting was nearly over, and certainly after the folding and thrusting had ceased, tensional phases developed and the strata were broken by large normal faults. Igneous intrusion also came in toward the close of this crustal disturbance and perhaps added to the structural deformation which resulted from the earlier folding and faulting. Evidence of added structural deformation by the intrusion is not so clear in Cassia County, but is marked farther north about the main Idaho batholith.

The crustal disturbance resulted in uplift of the region, while subsequent erosion carved the region into mountains that were probably higher and more rugged than those of the present day, as suggested from the great thickness of strata which has since been removed.

CENOZOIC ERA

After the production of the mountainous topography in the Mesozoic era, erosion continued to wear away the land and the subsequent history is mainly one of erosion interrupted from time to time by crustal disturbance and by the deposition of continental beds and lavas. Reduction of the Mesozoic mountains was apparently well advanced in the Eocene, for in parts of southeastern Idaho the Wasatch beds belonging to that period were laid down on a nearly peneplained surface. This surface has not been recognized in Cassia County unless the flat area on the top of Mount Harrison should bear it, but it may be represented in the mountains in northern Idaho by one of the extensive summit erosion surfaces in that region.[1] Erosion in Cassia County probably continued through the Oligocene and perhaps well into the Miocene. By Miocene time the land must have worn down to a surface of, old age or to a peneplain; such as now is retained on the resistant quartzites on the higher parts of the Albion Range. This surface is perhaps the same as the Snowdrift peneplain[2] described by Mansfield in some of the mountains in southeastern Idaho and perhaps the same as the subsummit erosion surface in the mountains of northern Idaho.[3]

Near the middle of the Miocene, or perhaps even earlier, active erosion was again revived as a result of marked crustal disturbance, chiefly one of broad regional uplift. This uplift may have been accompanied by normal faulting. As a result of the rejuvenation, broad and deep valleys were carved in the uplifted, old, worn-down surface. Erosion in the Middle Miocene eventually gave way to aggradation and these old valleys (Tygee erosion surface of Mansfield) were floored with extensive fluviatile and lacustrine deposits. These deposits continued to accumulate until all the valleys and some of the lower hills were blanketed. Then volcanism broke out on a grand scale and magmas found their way to the surface probably along normal fault planes that developed as a consequence and subsequent to the uplift. During the early period of the volcanic action, vast quantities of pumiceous ash showered the country and became intercalated in the upper part of the fluviatile and lacustrine deposits. Some of the ash or tuff was probably reworked by streams and redeposited along with the sedimentary clastic material; most of it probably remained as it fell. This increased the thickness of the Upper Miocene blanket by many hundreds of feet and probably buried most of the higher hills so that the area represented a vast aggraded plain. Toward the last came quiet outflows of latitic lavas and in some places a flow of basalt. Some of the flows were intercalated with the showers of ash and tuff, but toward the end of the epoch the explosive action began to decrease in intensity and

[1] Anderson, A. L., Cretaceous and Tertiary planation in northern Idaho: Jour. Geology, Vol. 37, 1929, pp. 747-764.
[2] Prof. Paper 152, p. 14.
[3] Anderson, A. L., Jour. Geol., Vol. 37, op. cit.

increasingly greater quantities of lava, mainly quartz latite, were poured out on the surface. Extrusions may have continued into the early Pliocene, but by the end of the epoch of aggradation and volcanism, the general region must have been a vast generally featureless lava plain covering many thousands of square miles and exceeding greatly the present areal extent of the Snake River Plains.

After these eruptions the region probably suffered further slight deformation, including some uplift probably accompanied by tilting and faulting of the lavas and underlying strata in some regions. This deformation was not nearly so violent as the preceding, but considerable relief resulted, for in southwestern Idaho Kirkham[1] finds that considerable erosion intervened before the Idaho Pliocene lake beds were deposited and these now rest on a surface of somewhat irregular relief. Erosion probably continued throughout much of the Pliocene in Cassia County, but for most of the time the land must have remained comparatively low. Further uplift came toward the end of the Pliocene and this one marked the major diastrophic event since the Laramide revolution. The region was further domed or uplifted, moderately folded and probably faulted. Active dissection of the region was again renewed and before the close of the Pliocene the land had again been greatly reduced and an old erosion surface developed. This surface is now retained on the highest ridges in the Sublett Range and probably is to be correlated with the Gannett erosion surface described by Mansfield in the region farther east.

The Quaternary history of the region is largely one of erosion interrupted by uplift and faulting and modified by climatic changes and volcanic outbursts. Uplift again occurred early in the Pleistocene and soon after another erosion surface had been carved below the Gannett surface. This one is best developed in the Sublett range and is probably to be correlated with the Elk Valley surface described by Mansfield. Uplift again inaugurated another cycle of erosion and relatively wide shallow valleys were carved in the Elk Valley surface. This cycle probably correlates with the Dry Fork cycle farther east. Again further uplift permitted canyons from 500 to 1,000 feet deep to be carved in the floor of the Dry Fork valleys. This records the last general uplift in the region, and is probably the same that inaugurated the Blackfoot cycle farther east. None of these surfaces, except perhaps those of the Blackfoot cycle, can positively be identified in the western part of the area, probably because faulting of the normal or block type has been so recent that earlier features have been either eliminated or so modi-

[1] Jour. Geology, Vol. 38, op. cit., pp. 652-659.

fied that they cannot be recognized. It is possible that faulting was renewed during each of the uplifts, but the major movement probably antedated the Blackfoot cycle. It is during the Pleistocene epoch that the tilted block ranges and the other present topographic features were produced. Recent erosion has accomplished little change since the Blackfoot cycle. Drainage was somewhat disturbed as the result of the crustal dislocations, but some of the major streams, such as Raft River and Cassia Creek, were able to continue in their courses across the Malta Range, although Marsh Creek was entirely turned aside by the faulting which produced the block range.

Possibly during the Pliocene, and surely in the Pleistocene, came the subsidence or downwarp of the Snake River Plains area. In western Idaho subsidence apparently began in the Pliocene and continued throughout much of the Pleistocene and into recent times, but so far as evidence in Cassia County is available, much of the downwarp has occurred since the early Pleistocene block faulting and the development of the Elk Valley and Dry Fork erosion surfaces. Subsidence had largely ceased, however, with the inauguration of the Blackfoot cycle, although as elsewhere in the general region some movement might still be in process. During the subsidence, and especially after most of the down-warping had taken place, vast floods of basalt issued from fissures or vents along reopened earlier fault planes and filled that great depression to its present level, building up the vast Snake River basalt plain. Some of the basalt flooded the edges of the structural valleys causing additional silting. Some streams were turned aside by the basalt as it encroached upon the region. Thus Raft River, Marsh Creek, and Goose Creek have each had to form new channels at the edge of the basalt plain.

In the Pleistocene epoch here as in other mountainous regions in the Northwest, glaciation was prevalent in the higher parts of the mountains. Extensive glaciation is recorded rather early in the Pleistocene, perhaps even before the Blackfoot cycle had started. At that time South Mountain and the Albion Range were extensively glaciated and it is possible that other ranges carried glaciers, but evidence of them is not retained in the less resistant rocks. Tongues of ice projected from the top to the very base of the Albion Range along much of its eastern side, carving the valleys into their broad U-shapes, which, because of the resistance of the quartzites to erosion, have not been notably altered to the present day. Glaciers also descended the east slopes of South Mountain and in the basin below formed a large piedmont or basin glacier from which tongues extended entirely across the Albion

Range to the basin at Almo and also northward along the basin between the two ranges to the edge of the alluvial plain near Oakley. Moraines deposited at the margins and ends of the glacial lobes have generally not survived erosion and aggradation, but have been smoothed over or buried under more recent piedmont alluvial deposits. Certain high terraces and dissected surfaces in the Black Pine Range suggest glacial origin. These are high above the present valley or canyon floors, and, if they represent this same epoch of glaciation, offer evidence of a pre-Blackfoot age for the glaciers.

Glaciation also occurred much later in the Pleistocene. This glaciation was much less extensive than the earlier and the glaciers were clustered on the north and east sides of only the highest peaks. Their lower limits, except on the steep easterly slopes of Cache Peak, did not pass below 7,000 feet A.T. These carved cirques and rock basins of notable size, in which lakes are still impounded. Morainal forms are fresh and but little changed by subsequent weathering and erosion. These relations suggest a comparatively recent or late Wisconsin stage of glaciation. These glaciers occurred in valleys belonging to the Blackfoot cycle. During this time and perhaps a little earlier, Lake Bonneville washed the southeast and east slopes of the Black Pine Range.

Filling of the structural basins and valleys has been an important process since the ranges were blocked into their present forms rather early in the Pleistocene. Waste from the higher border lands has strewn the basins and has aggraded them, raising their levels hundreds, and in some of the larger perhaps a thousand, feet or more. This process still continues, for the vast piedmont alluvial plains increase in size yearly as debris is brought down from the higher lands.

MINERAL RESOURCES

Cassia County has a considerable variety of mineral resources, including metallic but especially nonmetallic varieties. These resources have, however, been but little developed and constitute in the main large potential reserves.

Among the metallic resources are deposits of lead-silver ores, zinc-silver ores, and inconsequential amounts of other metals. The non-metallic resources offer much greater variety and among them may be listed building stone, limestone, road metal, quartzite, volcanic ash, mica, clay, feldspar, and cyanite. There is also the possibility that the Phosphoria formation, which contains immense quantities of phosphatic shales in southeastern Idaho and constitutes the chief mineral resource of that region, might contain the phosphatic shale horizon here. Water for irrigation is the most valuable resource in the County, as nearly the entire population of the County is directly or indirectly dependent on it for a livelihood. Neither the water resources nor the phosphate possibilities will be discussed in this report, but because of their magnitude should be made the basis for separate detailed investigation.

ORE DEPOSITS

HISTORICAL SKETCH

It is not known just when the first discovery of ore was made within the district, but it was probably rather late in the past century. Several locations were made about the year 1880, among these the Silver Hills mine on the Black Pine Range, and several properties on Fairchild and Conner Creeks in the Albion Range. As most of the properties have changed hands several times, it is exceedingly difficult to trace through their histories. Most of the properties were probably discovered and located about the same time, and each worked in a desultory manner for many years after. Only within the past two decades have attempts been made to work any of the deposits by well organized companies. Most important of these are the Silver Hills Mining Company, which controls 30 unpatented mining claims in the Black Pine District, among them the Silver Hills mine; and the Melcher Mining and Milling Company, operating several claims in the Stokes district on Conner Creek in the Albion Range. The Melcher Mining and Milling Company was incorporated as far back as 1899 and has done much development, including the installation of a modern milling plant. It, however, has been inactive during most of the past decade. The Silver Hills Mining Company was incorporated in 1920 and has continued more or less active to the present day.

Sporadic shipments of ore have been made from the district since the early discoveries. It is reported that some ore was shipped from the Silver Hills mine about 1894, also 14 cars of ore from the nearby Hazel Pine about 1914, and six cars of rich zinc ore from the adjoining Ruth property during the World War. Shipments of small amounts of ore are also reported from several properties in the Albion Range prior to 1900 and some shipments have been made since that time. Unfortunately the writer was unable to obtain any information from officials of the Melcher Mining and Milling Company and nothing is therefore known of its production or years of operation. So far as known no shipments have been made from any of the properties during the past decade, although prospecting has remained fairly active.

GEOGRAPHIC AND GEOLOGIC DISTRIBUTION

The ore deposits occur in two widely separated parts of Cassia County, one on the southeast slope of the Black Pine Range, to which the name Black Pine district has been given, and the other in the Albion Range. Deposits in the Black Pine district are grouped in a comparatively small area, but those in the Albion Range are much more widely scattered. Most of the mineralization in the latter is in the north section of the Albion Range, mainly on Conner Creek and to a less extent on Fairchild and Boulder creeks on the west slope of the range east and northeast of Oakley. Some mineralization also shows at the north end of the Range, not far from Albion and Declo. All of these deposits are assigned to the Stokes Mining district.

Only the older rocks of Paleozoic and pre-Cambrian age contain known mineral deposits. In the Black Pine district the deposits are in the Carboniferous rocks, mainly in the Brazer limestone. In the Stokes district they are in the pre-Cambrian Harrison quartzites and to a less extent in the Carboniferous rocks.

GENERAL CHARACTER OF THE DEPOSITS

The deposits are essentially of two kinds, fissure and replacement veins. This character is determined mainly by the nature of the enclosing walls. Veins in the quartzites of the Harrison series are essentially fillings in fractures, those in the limestones have formed by replacement directed along fissures or fractures.

The most important metals contained in them are silver, lead, and zinc. Other metals present in minor amounts are copper, gold, antimony, and mercury. Classed as to metals the veins are those of silver-lead, silver-zinc, zinc, and mercury.

Deposits in the Albion Range are mesothermal, according to Lindgren's scheme of classification;[1] that is, were formed under conditions of moderate temperatures and at moderate depths below the earth's surface. Those in the Black Pine district are epithermal; that is, formed at low temperatures and pressures and supposedly at but relatively slight depth below the surface. This genetic distinction is apparently the result of two distinct and widely separated epochs of

---

[1] Lindgren, Waldemar, Mineral deposits: McGraw-Hill Book Company, 1928.

metallization.   The mineralization in the two districts is therefore wholly different and unrelated.

MINERALIZATION IN THE STOKES DISTRICT

GEOGRAPHIC FEATURES

Mineralization in the Albion Range or the Stokes district is confined mainly to three rather widely separated localities.   The most extensive mineralization has been on Conner Creek on the east side of the range, about seven miles due south of the town of Albion or twice that distance by road.   Most of the deposits are on the north side of the valley and at the head of the creek.   The most extensive workings are in this part of the district, and among them are the properties of the Melcher Mining and Milling Company and the Big Bertha Mining Company. Several others have also been extensively developed or prospected. The property of the Melcher Mining and Milling Company is the only one accessible by automobile.   All properties lie high on the sides of Conner Creek valley at elevations ranging from 7,000 to 8,200 feet A.T. The Old Bertha is on the summit erosion surface at the top of the range, overlooking Conner Creek valley.

Mineralization has also been fairly widespread on the west slope of the range, mainly along Fairchild Creek and Boulder Creek.   These places are easily reached from Oakley.   Much of this region has been well prospected, but no deposits of consequence have been developed. This part of the district lies seven miles and more from the deposits on Conner Creek.

There is occurrence of mineral also in the northern end of the Albion Range.   The Old Dominion property lies about four miles northeast of the town of Albion.   It is the only mining property in this part of the range.   Minor occurrences of mineral are reported nearby and particularly near the extreme end of the range east of Declo.   This part of the district is easily accessible.

GEOLOGIC FEATURES

Deposits along Conner Creek are in the quartzites and micaceous members of the lower division of the Harrison series.   These are centered very near the crest or axis of the dome or anticline which constitutes the main structural feature of the northern section of the range. Those lower on the creek are on the east flank of the anticline and those near the head of the valley are on the west side.   Some of those between are very likely along the axis.   Specific grouping has been, however, largely about a small stock of porphyritic granite whose intrusion has been near the center of the dome or near the axial line of the dome. Deposits lie near the east, west, and north margins of this stock and in part within its borders. This grouping is significant as it suggests that the mineralization has a genetic connection with the granitic intrusive.

On the west slope of the range the deposits occur in part in the middle division of the Harrison series and in part in the underlying block of Pennsylvanian strata. These deposits lie on the far west side of the dome and mainly in the overridden younger rocks. The strata both above and below the overthrust fault plane have been considerably disturbed, especially by subsidiary fractures.

The geologic relations at the north end of the range are very like those on Fairchild and Boulder creeks, except that the rocks may be even more disturbed by faults.

CHARACTER OF THE DEPOSITS

Deposits in the quartzitic members of the Harrison series occur essentially as fissure fillings, enlarged slightly by replacement. Those in the limestones are replacements. Most of the veins fall into the first grouping. Metals furnished by both kinds of veins are mainly lead and silver, and to a much less extent, gold and copper. Quartz constitutes the chief gangue mineral and generally the main filling. The veins may then in general be classed as silver-lead-quartz veins.

MINERALOGY

GENERAL SUMMARY

The main ore minerals of the veins consist of galena and to a less extent tetrahedrite, pyrite, and chalcopyrite. Arsenopyrite, specularite, and sphalerite are recorded, but are rare. Secondary minerals, though not profitably abundant, include a long list. Most common of these are pyromorphite, cerussite, anglesite, limonite, azurite, malachite, chrysocolla, and covellite.

PRIMARY MINERALS

Descriptions of the minerals will not be given in order of abundance or other distinctive features, but in the order of their deposition from the vein solutions. The listing is therefore paragenetic.

Quartz: Quartz is the most abundant mineral in the veins. Much of it occurs in massive form, less often as drusy crystals in clefts. It occurs mainly as a massive fissure filling, but locally it may replace schist or granite walls. Some occurs in reticulating veinlets in crystalline limestone, as in the Old Dominion deposit, or shows massive replacement of the limestone. The massive vein quartz has a white granular appearance and is moderately coarse.

Shattering or fracturing of the quartz has everywhere preceded the introduction of sulphides. Where fracturing of the quartz filling has not occurred the vein is barren, where fractured, sulphides have entered and have not only closed the fractures but have also enlarged them by replacement. The sulphides thus occur as tiny reticulating veinlets in the quartz, or, where replacement has progressed to advanced stages, as larger granules or bands of irregular size and width. In places where open clefts remained after the quartz had been deposited, the sulphides coat the quartz crystals. Some quartz presents good crystal outline against the sulphides, even in areas of replacement, and it is therefore likely that its deposition also continued in later or sulphide stages, but in more subordinate amounts.

Specularite: Specularite or specular hematite was noted in but one specimen from the Old Bertha. It occurs in tiny black scales (reddish streak) in a fracture in the white vein quartz. It is not in contact with sulphides.

Pyrite: Pyrite is not notably conspicuous in any of the deposits, although always present. It is apparently nonauriferous, for what gold there is in the veins seems to be associated with the galena.

Some of the pyrite is aligned along fractures in the quartz. In such position it always has excellent cubical outline, which it assumed during replacement of the quartz. Most of it, however, has been engulfed in later sulphides and against them its outlines are partly irregular, due to its partial replacement by them. It also forms small residual grains in the other sulphides.

Arsenopyrite: Arsenopyrite was noted only in ore from the Albion group in which it is sparsely associated with other sulphides. Its relation to pyrite is not known, although probably somewhat later in its deposition, but it does show engulfment by other sulphides and also partial replacement by them. Crystals or grains are microscopic in size.

Sphalerite: Sphalerite was noted in very spare amounts in ore from the Old Bertha mine, and also in ore from the Albion group. Its presence was determined only from microscopic examination. Some of it holds inclusions of pyrite. In turn it is similarly included and partly replaced by other sulphides. Most of it contains minute scattered grains or blebs of chalcopyrite arranged symmetrically along its crystallographic partings, an arrangement and relation usually ascribed to unmixing from a solid solution.

Tetrahedrite: Some of the silver content of the ore is probably due to the association of tetrahedrite with galena. Tetrahedrite occurs most commonly as microscopic inclusions in galena. More rarely it forms larger grains and masses. In the latter form at the Old Bertha it shows replacement of shattered sphalerite. In turn it is partly cemented and replaced by galena. In all the deposits it is mainly older than the sulphides not yet described.

Chalcopyrite: Chalcopyrite is nowhere abundant except in parts of the Albion vein. It usually forms as small grains in galena, or veinlets or granules replacing quartz, sphalerite, and tetrahedrite. Its deposition mainly preceded that of galena, for the latter shows some slight replacement structures.

Galena: Galena as the chief ore mineral forms grains, granules, seams, or bands in the vein quartz, usually with much less noticeable pyrite and chalcopyrite. Much of it tends to occur in massive, coarse-grained aggregates of cubical crystals, but that at the Old Dominion is very fine-grained and that in the Albion vein is mashed and schistose, due to post-mineral slippage along the vein. An assay on some nearly pure galena from the bins of the Melcher Mining and Milling Company shows values of 99.4 ounces in silver and 0.01 ounces in gold. Examination of polished surfaces of this ore reveals tetrahedrite as minute inclusions, but probably not sufficiently abundant to account for all the silver in the ore. It is probable therefore that the galena is argentiferous and carries some silver mineral in solid solution.

The galena is the last mineral to be deposited. It also cements the fractured quartz, replaces the quartz pieces, and cements and corrodes the other sulphides. At the Old Bertha it fills fractures in the chalcopyrite; in the Albion ore it partly replaces the chalcopyrite.

Gold: It is not known in what form the gold occurs in the ores. None of it was seen in the free state. As mentioned earlier, the gold values seem to increase with the lead content.

SECONDARY MINERALS

Copper sulphides: Covellite, and to less extent chalcocite, constitute the only secondary sulphides. Neither occur in sufficient abundance to be noted except in microscopic examination of the partly oxidized sulphide ore. Covellite generally forms patches in anglesite near galena residuals or as irregular borders on the galena. It also forms on borders of chalcopyrite in contact with galena. Elsewhere oxidation of chalcopyrite yields chalcocite.

Copper carbonates: Malachite, and less commonly azurite, form thin patches or coatings on the vein quartz or in the cavities left from the oxidation and removal of sulphides. They also form on the surface of the partially oxidized sulphide residuals. Neither can be considered as more than surficial traces in most of the veins.

Copper silicate: Chrysocolla and also copper pitch ore occur in some of the deposits. The copper pitch ore is but a minor product following the replacement of the chalcopyrite by chalcocite. Chrysocolla is confined mainly to the outcrop where it forms minor crusts and patches, usually with malachite and azurite.

Pyromorphite: Pyromorphite, the chlor-phosphate of lead, is conspicuous as a secondary lead mineral near or at the surface of most of the lead veins and usually in the surrounding country rock. It occurs mainly as greenish crusts or coatings on the vein quartz or quartzite, or as crystalline aggregates lining clefts and fractures. Its crystals are mainly prismatic or barrel-shaped, less commonly fibrous. In its non-crystalline occurrences it assumes reniform or botryoidal forms. Colors range from deep green to yellow-green and less commonly to brown. The mineral is particularly noticeable at the surface along the veins near Conner Creek.

Cerussite: Cerussite, lead carbonate, is common in the upper part of most of the veins, usually as ashy gray crusts or in earthy forms, more rarely in crystals. Unfortunately, oxidation has nowhere extended far beneath the surface and the carbonate zone is not thick. Much of the cerussite also encloses residua of galena. Some, along with the partly oxidized galena, has been marketed.

Anglesite: Anglesite, the sulphate of lead, is conspicuous only as an incipient oxidation product of galena. Its usual occurrence is as minute veinlets in galena, developed by replacement and directed by cleavages within the galena. Anglesite is not a particularly stable form of lead, but the sulphate is usually replaced by the more insoluble carbonate.

Iron oxides: Iron gossans are not conspicuous except on the high ridge at the head of Fairchild Creek. The gossan is composed mainly of limonite, but has a little hematite and a little manganese oxide. Its formation is probably the result of the oxidation of pyritic bodies. Assays for gold and silver revealed the presence of neither.

Some iron oxides occur in the upper parts of the veins on Conner Creek and elsewhere, but the amount is not great, probably because neither pyrite nor chalcopyrite are abundant as primary minerals.

MINERAL PARAGENESIS

Clear evidence can be obtained from polished sections that crystallization did not take place simultaneously, but that the different primary minerals were formed successively. The intimate intermingling of the minerals show convincingly that the great mass of the ore in most of the veins was formed from solutions of essentially the same character that underwent gradual change instead of from a succession of distinctly different solutions. Deposition was sufficiently interrupted at times by movement along the vein to shatter or fracture early minerals. Most pronounced of these movements concurrent with mineral deposition occurred after most of the quartz had been deposited and before the sulphides were deposited.

The paragenetic sequence, without restating the essential relationships brought out in the description of the individual minerals, may be given as follows: *quartz*, (specularite), *pyrite*, (arsenopyrite), sphalerite, *tetrahedrite, chalcopyrite*, and *galena*. Most abundant or most valuable minerals are italicized, those whose positions are not accurately fixed are in parentheses. This sequence exhibits no unusual peculiarities, but is rather characteristic of quartz-lead veins in the Cordilleran region in or near intrusive bodies of granitic rock.

DEPOSITS

STRUCTURAL FEATURES

Most of the veins occupy well-defined fissures whose walls are sharp and distinctly set off from the vein filling. Less commonly the deposits constitute small quartz stringers in poorly defined shear zones, and this kind is only slightly less common than the irregular bodies developed by replacement of calcareous rock. Quartzite, especially, has not been amenable to replacement and fissures in such rock have had their essential forms retained. Schist and granite have been less resistant to replacement by the mineralizing solutions and fissures in them show some modification by wall replacement.

Fissure veins are typically tabular, but exhibit a variable amount of pinching and swelling. Locally some of the lenticular bodies attain chimney-like proportions near fracture intersections, but in general the veins seldom exceed six feet in thickness and usually not more than three. Most of them are persistent for several hundreds of feet, and extend to considerable depth, although in no place has development been sufficient to afford a satisfactory basis for estimating the actual depths to which the deposits might be expected to extend. Deposits in calcareous rock on the other hand usually swell into large chimney-like masses as much as 50 feet across. This kind is very erratic and occurs only infrequently along the zone of fissuring and may pinch out in depth as sharply as it does horizontally. Much of the chimney mass consists of a ramifying network of small quartz stringers. Some stringers also occur in sandstones and quartzites, locally forming a reticulating vein system. Individual stringers usually do not exceed an inch in thickness and are too widely spaced to form profitable deposits.

Position of the fissures with respect to the major structural features of the region has already been cited. Those on Conner Creek are near or essentially along the anticlinal axis of the range; those elsewhere, on the far west flank of the anticline or in the underlying block of Carboniferous rocks. In general the direction of fissuring closely parallels the trend of the major structures, and the individual fissures strike mainly to the northeast from 10° to 45°, more rarely to the northwest. The dip of the fissures is usually steeply to the west or northwest.

None of the mineralized fissures give any indication of large relative displacement. No doubt these fissures are genetically related to the major structural disturbance in the region and antedated the intrusion of the granitic bodies. None of the larger faults, which are believed to have been produced at this time, either normal or reverse, are known to be mineralzed, perhaps because the gouge along such faults was so finely comminuted and so tightly compressed that it was impervious to the passage of ore solutions. Several of the veins are in part in granitic rocks, and the fissuring preceding their formation obviously took place after the intrusion and consolidation of the rock. The major orogenic movements and the batholithic intrusion in this part of Idaho are probably intricately related parts of the same general process. Part of the disturbance of the stratified rocks at any particular place doubtless preceded the advent of the granitic magma, and part of it took place concomitantly with and shortly after the intrusion. It seems probable that the fissuring which provided paths for the ore solutions was a minor feature of this great process of mountain building and intrusion and that it continued without interruptions throughout the period of disturbance.

There is abundant evidence of recurrent movement along the fissures during the mineral deposition and after the major period of ore deposition. In some of them minerals first deposited were intricately brecciated before the later minerals were formed and this process apparently took place several times and to a varying extent in all the deposits. Some of the ore, especially the galena ore at the Albion vein, has been extensively crushed and ground into a fine-grained aggregate, showing a schistose structure. This feature is due to post-mineral movement. Post-mineral faults crossing the veins have been found in the course of underground development, but displacement along these has not been great.

VEIN STRUCTURE

Most of the fissures have been entirely filled with massive quartz throughout the greater part of their length, but here and there open clefts remain filled by drusy quartz crystals. Occurrence of open spaces is the exception, however, rather than the rule. None of the veins show evidence of crustification other than the occasional open cleft, nor do they show evidence of banding. Sulphides, and in particular galena, occur as small grains, granules, or masses scattered irregularly through the vein quartz, usually directed along small fractures in the quartz. Less commonly the galena forms massive bands or seams, then usually not more than two or three inches wide and of no great length. Such massive bands are usually lenticular. In some places the galena forms irregular masses as much as six inches in diameter, but mostly its occurrence is in much smaller masses of such size that milling is an essential process for its separation from the quartz gangue.

The veins are seldom mineralized with ore for more than two or three hundred feet along the strike and usually for much shorter distances. Ore shoots have been in part formed along the shattered zones in the early quartz filling. There is no way, so far as the writer knows, of forecasting the occurrence of such zones or their disappearance. Localization of ore shoots along some of the veins has also in part been determined by cross fractures or fissuring. Increased porosity resulting from such fracture intersections seems most favorable to the passage of mineralizing solutions and the development of ore shoots.

SECONDARY ENRICHMENT

Ore from near the surface, recovered in the early days, was mainly oxidized and consisted largely of cerussite enclosing residual grains and lumps of galena. This ore was probably richer than that at greater depths, for shipping ceased soon after the essentially primary ore had been encountered. Part of its enrichment was in silver, but no information concerning the silver and lead ratios could be obtained. This secondary ore is no longer an important factor and does not warrant further discussion.

In some of the veins enrichment or even oxidation is scarcely noticeable more than a dozen feet below the surface. In others, partially altered ore extends from 50 to 100 feet below the surface, but throughout this depth there has been little enrichment.

GENETIC CLASSIFICATION

Deposits such as these in the Stokes district fall into the group regarded by Lindgren[1] as mesothermal; that is, into those metalliferous deposits formed at intermediate temperatures by ascending thermal solutions and in genetic connection with intrusive rocks. Absence of such minerals as magnetite, garnet, pyroxene, tourmaline, pyrrhotite, shows that a high degree of heat did not prevail at the time of genesis, but that the temperature must have ranged from 175° to 300°C. The thin seam of specularite found in one of the veins might suggest, locally at least, that temperatures were somewhat higher; but specularite is not a reliable thermal indicator, for it is known to occur in deposits of low temperature characteristics. But in most deposits in the Stokes district the conditions were much below the hypothermal range, although still much above the epithermal. That conditions of deposition were not epithermal or near surface is indicated from the complete absence of minerals characteristic of that type of deposit. The deposits in the Stokes district also lack the extremely brecciated structure common in deposits formed near the surface and also the lenticular form of deep-seated deposits. Fissures are fairly regular in strike and dip and apparently stood partly open before being filled with vein matter. There is no way of determining the exact or even approximate depth of these deposits beneath the surface at the time of their formation, but it probably exceeded a mile. Minerals in them, such as the association of pyrite, galena, sphalerite, chalcopyrite, tetrahedrite, arsenopyrite, are especially common in the mesothermal type of deposit.

AGE AND GENESIS

It has come to be generally recognized that ore deposits such as those here described are formed by heated solutions derived from intrusive igneous rocks. There can be little doubt that the mineralization

[1] Lindgren, Waldemar, Mineral deposits: McGraw-Hill Book Co., pp. 598-717.

in this district is related to the Cassia batholith and its outliers. Arrangement of the veins about and in the upper part of the stock on Conner Creek is especially suggestive of such a relationship. It has already been shown that the fissures were produced before the advent of the mineralizing solutions, and the solutions followed the fissures because they offered the least resistance to their passage. It has also been explained that most of the fissuring occurred during the great orogenic movement when the strata were folded and broken by faults, and that intrusions of granitic magma followed soon after. As the mountain building is believed to have taken place in the late Cretaceous and to have extended into the early Eocene and as the batholithic intrusions came during the later part of this epoch, the mineralization is therefore late Mesozoic or more likely early Tertiary. The characteristics of these deposits differ sharply from those in the Black Pine district and from those in the Miocene lavas in neighboring or nearby districts,[1] where the mineralization is related to a younger epoch of igneous activity. They resemble to some extent deposits associated with granitic rocks of similar age elsewhere in Idaho[2] and do not resemble deposits associated with Tertiary rocks either in Idaho or in other parts of the Rocky Mountain region.

COMPARISON WITH DEPOSITS IN OTHER REGIONS

Prominent veins carrying milky quartz and sparsely disseminated tetrahedrite, galena, and usually zinc blende, with subordinate pyrite, are common in the Cordilleran region of the United States in or near intrusive bodies of granitic texture. According to Lindgren,[3] many such veins are found in the great batholiths of Idaho and Montana, but the deposits are on the whole poor and rarely worked, although from 1870 to 1890 the enriched surface zones in many places yielded much silver. These veins are not to be confused with the rich galena-siderite veins in the famous Coeur d'Alene district of Idaho or with the formerly productive tetrahedrite-galena-siderite veins in the Wood River district of Idaho. The galena-quartz veins such as are represented in the Albion Range are essentially distinctive from them, although probably formed at the same time and under similar conditions, but from solutions of somewhat different composition, especially from solutions low or deficient in iron and carbonate.

---

1 Anderson, A. L., Geology and ore deposits of the Lava Creek district: Idaho Bureau of Mines and Geology Pamphlet 32, 1929. Ross, C. P., Ore deposits in Tertiary lava in the Salmon River Mountains, Idaho: Idaho Bureau of Mines and Geology Pamphlet 25, 1927.

2 Ransome, F. L., and Calkins, F. C., The geology and ore deposits of the Coeur d'Alene Mining district: U.S. Geol. Survey Prof. Paper 62, 1908. Umpleby, J. B., Westgate, L. C., and Ross, C. P., Geology and ore deposits of the Wood River region, Idaho: U. S. Geol. Survey Bull. 814, 1930.

3 Mineral deposits, op. cit., p. 641.

The veins in the Stokes district are perhaps closely allied to the galena-quartz veins in the central and eastern Cordilleran states, especially those described near Kingman, in northwestern Arizona, in the Wallapai[1] mining district. These deposits have been valued mainly for their rich secondary silver ores. The veins occupy well defined fissures, with steep dip, and have defined walls. The gangue is quartz, in places shattered and cemented by a later generation of calcite and occasionally siderite. Among the primary sulphides are pyrite, arsenopyrite, galena, zinc blende, and chalcopyrite, tennantite, proustite, and pearceite. Veins are narrow, are fissure fillings, and their structure is irregularly massive. Unfortunately the veins in the Stokes district have not been as well mineralized as those elsewhere.

OUTLOOK FOR THE DISTRICT

As a whole galena-quartz veins such as are found in the Stokes district are poor and rarely worked after the enriched portions near the surface have been removed. Some of the veins in the Stokes district have small, moderately well mineralized ore shoots, but most of the veins are not especially well mineralized and judging from the long period of inactivity other than ordinary annual labor even over periods when lead commanded high prices, mining has not always been a profitable venture. Tetrahedrite is usually a very minor mineral in the veins and does not increase the tenor of the ore to any great extent. The veins must therefore be mined for their lead content and for what silver the lead contains. Not until the price of lead is well above the prevailing price in 1930 will the outlook for the district be other than pessimistic. Gold apparently does not occur in sufficient quantity to raise the tenor of the ore to a level that would make exploitation profitable.

PROPERTIES

MELCHER MINE

The Melcher mine lies on the north side of Conner Creek valley in Sec. 8, T. 13 S., R. 25 E., at an elevation of about 7,000 feet A.T. It is easily reached over a road from the Elba Basin, the lower part steep and rocky but the upper part, though rocky, on an even grade. Since 1899 the mine has been controlled by the Melcher Mining and Milling Company, which not only undertook extensive development of the deposit but equipped the property with a complete mill and mine camp. The mill is modern and is equipped with four flotation cells with vertical

---

[1] Schrader, F. C., Mineral deposits of the Cerbat Range, Black Mountain, etc.: U.S. Geol. Survey Bull 397, 1909; and Bastin, E. S., Bull. 750, U.S. Geol. Survey, 1924, pp. 17-39.

electric motors, filter press, cone classifier, ball mill, jaw crusher, and three Wilfley tables. Electric power carried over high tension lines from Albion was formerly used, but later a steam driven plant was installed. It was not ascertained just when the mill was erected, but it has remained idle at least since 1920. The mill and camp lie about 700 feet below the mine tunnels and ore is lowered to the mill over a tram. Underground workings consist of two main tunnels, one about 3,300 feet long, the other 1,600 feet long, and several shorter tunnels, shallow shafts, and cuts.

No information was obtained dealing with the early history of the property or of its more recent operations. The mill was apparently operated at several intervals and considerable concentrate marketed. These concentrates consisted chiefly of galena, but are reported to carry about 4 per cent copper, some values in silver, and about $10.00 in gold per ton. This property was idle in 1930 and had been idle since 1920. Efforts to get in touch with the owners of the property and to obtain from them information concerning the past history, past production, and tenor of the ore were unavailing. Some of the tunnels were caved, and as there was no one on the property at the time of the writer's visit the results of the examination were far from satisfactory.

Veins on the property are of the fissure type and consist essentially of a quartz filling with irregularly scattered masses of galena and less frequently pyrite and chalcopyrite. Tetrahedrite is revealed in small grains in the galena on microscopic examination. Most of the galena is coarse cubical, but a little has been mashed to form steel galena. At the surface or in the upper workings small patches of malachite and azurite occasionally crust the quartz or ore minerals. Greenish crystals and crusts of pyromorphite may also be seen on the surface croppings and in some places in the adjoining country rock. Little cerussite and anglesite was observed, although both may have been present in some abundance before mining activities began. Scarcity of secondary minerals is undoubtedly due to the Pleistocene glaciers, which have cut deeply into the valley and removed any previous oxidized or enriched zone that might have existed.

Several veins occur on the property but only one has been extensively developed. This one is mainly in the quartzite but occasionally cuts schist and granite. Others have similar relations, although several of the minor ones are almost entirely in the granite. The granite does into the roof of the stock. A short distance below the surface it becomes the prevailing kind and the host for most of the veins. These veins are on the east side of the anticline, although not far from the axis, and are in and above the roof of the granitic stock mentioned above. The principal vein at the surface strikes N. 30° E., and dips 55° N.W. It shows some slight changes in both strike and dip below the surface and also some offsets because of post-mineral faulting. At the surface its width varies from 18 to 20 inches, but underground it increases in places to six feet. Some of the other veins have similar trends, but several strike to the northwest.

The principal vein has been opened by two main tunnels and by several shorter ones. These are joined by raises and shafts, some of which extend to the surface. The lowest tunnel is not entirely accessible, as ore from a broken bin well back from the portal has effectively sealed off the remainder. The next tunnel, about 100 feet above, is caved at the portal. The conditions of the raises between the two tunnels did not encourage examination of the vein in the higher tunnel. Only about 910 feet of the lower (Grant) tunnel could be examined. It is reported to continue for at least another 2,000 feet. From the blocked point to the portal·the vein alternately pinches and swells, generally from nothing to three and one-half feet, or in some places breaks up into smaller stringers. One of the lenticular masses contains scattered masses of ore. This character is maintained for about 145 feet, the fissure showing some change in curvature as well, and then the vein enlarges to six feet and maintains this remarkably uniform width for about 250 feet. Much of this part of the vein carried ore, as evidenced from the stoped area above. The fissure then crosses soft schist, and, although the fracturing is fairly pronounced, the greater quantity of gouge developed did not permit ready access to the mineralizing solutions and the fissure is mainly barren, with recurrent small lenticular quartz masses. Much drifting has been carried to the west from a point 300 feet from the portal, but only small seams of quartz have been disclosed. This drift also reaches the surface and it represents the earliest work, subsequently shunted off by a shorter and more direct tunnel to the vein. The tunnel about 100 feet above is reported to be but 300 feet long. Underground the vein has an average trend of N. 20° E., but varies from N. 10° E. to N. 30° E.

Several cuts have been made higher on the vein, and numerous cuts and shallow shafts have been sunk on other veins to the east, including several tunnels now inaccessible. Small quantities of galena and pyrite are strewn on some of the dumps. One vein has an especially favorable outcrop, but a long crosscut driven far below the cropping revealed nothing with depth. Although the vein is in quartzite at the surface, the tunnel below was almost entirely in igneous rock. The tunnel is reported to be 1,600 feet long. It is caved at the portal. Quartz stringers and veins are numerous in the valley slope, but few of them are more than lightly mineralized.

BIG BERTHA GROUP

The Big Bertha group lies at the summit of the wide flat ridge between Conner Creek and Howell Creek at an elevation of about 8,200 feet A.T., about two miles northwest of the Melcher mine. It cannot be reached by car, but supplies and equipment have been brought in by wagon. This property, consisting of 13 unpatented claims, is owned by the Big Bertha Mining Company, incorporated in 1924. It is reported that 12 tons of ore were shipped from the property about 40 years ago, consisting mainly of lead carbonate with good values in lead and silver. Last work on the property, other than annual labor, was in 1926.

Little could be learned on the ground, as prospecting and development has had to proceed through shafts, all of which are now inaccessible. Several shallow cuts, however, give some information concerning the veins and mineralogy. Judging from the alignment of the cuts and shafts there must be several veins, all of which have yielded a little ore. These veins, where they can be examined at the surface, are no more than a foot wide and consist essentially of milky quartz with sporadic recurring small pockets of galena. At the surface most of the galena has altered to cerussite and to lesser extent to pyromorphite. Other minerals identified in polished surfaces consist very subordinately of sphalerite, tetrahedrite, chalcopyrite, arsenopyrite, and specularite. Assays on ore samples (possibly concentrates) are also reported to carry $14.00 to $16.00 in gold per ton. As the shafts and cuts are aligned for several hundred yards, it is apparent that the veins must be of equal length.

One of the shafts is 120 feet deep, another 80 feet. Several others are from 35 to 40 feet deep. From one several short crosscuts have been driven. One long crosscut has been started some distance below the crest of the ridge, but it has not reached the veins, nor has another tunnel started nearby and abandoned. These are reported to be 500 and 400 feet long respectively.

ALBION GROUP

The Albion group, formerly known as the Badger, lies about a mile west of the Melcher mine and at about the same elevation. It is reported that about five carloads of ore have been shipped, one car of

30 tons returning values of $2,200.00 in lead, silver, and gold. Most of the work consists of a tunnel about 600 feet long, now caved about 470 feet from the portal.

The fissuring trends about N. 12° E. and varies in dip from vertical to steeply west. Most of it passes through schist whose strike is N. 50° W. and dip 25° SE., but it also extends across granitic apophyses. Most of the fissure is but very sporadically mineralized, and a series of disconnected lenses constitute the vein, usually not more than three feet across at the widest, together with minor seams and stringers. Deposition of the quartz in these lenses has been mainly by filling, also in part by replacement of the schistose wall. In some places the main fissure is intersected by other fissures, and these, too, carry narrow quartz veins.

Mineralization consists mainly of galena, much of it granulated or mashed by post-mineral movement along the plane of the vein, and of lesser pyrite and chalcopyrite, and of sparse tetrahedrite and sphalerite. Most of the sulphides form narrow compact seams or bands in the vein quartz, although some is disseminated. The massive bands are seldom over two inches wide, but some do attain a width of a foot or more. Such bands are always short. Chalcopyrite is more abundant in this vein than elsewhere in the district and seems to increase in quantity with depth. Clefts in the vein lined with drusy quartz crystals are also common.

These lenses have been followed for more than 470 feet, but the stoped area does not cover much over 200 feet. Small raises or stopes are numerous, and from their shapes indicate that the shoots must be more or less chimney-like, although not more than four or five feet wide and usually not more than twice as long. The stopes follow only the richer part of the shoots and are not continuous above the tunnel level. Stoping has not extended upward for more than 50 feet. Depth gained by the tunnel is about 150 feet.

At the surface the vein shows greater width, nearly eight feet. In one place it consists of 30 feet of barren quartz. Several shafts and cuts have been made on the outcrop higher on the ridge, also a 100-foot tunnel has been driven on the vein. Some malachite and considerable pyromorphite show at the surface.

GOLDEN EAGLE PROPERTY

The Golden Eagle property lies about a mile west of the Albion group or high above the sharp bend in Conner Creek at an elevation near 7,200 feet A.T. About 10 tons of ore were shipped in the early days of mining, but there has been no recent production. Workings are old and generally inaccessible. One shaft is reported to be 140 feet deep.

The vein is of the fissure type in quartzite and strikes about N. 45° E. and dips 70° NW. It has been explored by several cuts and winzes from the surface for a distance of about 50 yards along the strike. The vein is from two to four feet wide and contains mainly oxidized ore, consisting largely of cerussite with residual grains or kernels of galena, also of a small amount of pyromorphite. Assays made on some of the carbonate ore showed 0.04 ounces in gold and 15.6 ounces in silver per ton.

A tunnel has been started about 150 feet below the outcrop and several hundred feet to the east, and, although it has passed through about 300 feet of rock, it has not cut the vein. There is no way of ascertaining the character of the mineralization or the size of the vein with depth.

HARRIS AND HUME PROPERTY

The Harris and Hume property lies near the head of Conner Creek at an elevation of about 7,000 feet A.T., about two miles west of the Melcher mine and one-half mile north of the Golden Eagle.

The country rock consists of white quartzite, in part the conglomerate gneiss facies of the lower division of the Harrison series. These strata strike N. 40° W. and dip 15° SW.

Material on the dump of the main tunnel consists largely of quartzite fragments, together with some ore and vein quartz, in part drusy. Some of the quartz is coated with pyromorphite and most of it is stained brownish and black with limonitic varnish. The tunnel contains about 750 feet of working, including two crosscuts, one 35 feet long and the other 25 feet long. About 545 feet of the tunnel is on a very pronounced fissured zone trending N. 10° E. and dipping 70° NW. The quartzite along the fissure is intensely shattered and the rock reduced to small sharp fragments, even pulverized in places. This shattering forms a zone from two to 12 feet wide. In several places it contains small lenses of vein quartz, usually but a few inches wide, but one locally swells to four feet. These seams have no ore minerals. Above is a short tunnel and shaft. The vein near the shaft has some pyromorphite and limonite and it is probably from this place that the ore seen on the dump of the lower tunnel was obtained.

OLD DOMINION MINE

The Old Dominion mine lies about four miles northeast of Albion in the north end of the Albion Range in Sec. 16, T. 11 S., R. 25 E. The property was located in 1888 and produced a little ore in the several succeeding years. Workings are not extensive, but include several cuts and shafts, also three tunnels, two of them about 175 feet long and one about 50 feet long.

Several veins have been prospected, but only one has given encouragement to development. All are inclosed in limestone or marble and are replacement deposits along zones of fissuring. Pre-Cambrian quartzite overlies the marbleized limestone a short distance above, but it is not known whether the marble belongs to the pre-Cambrian series or represents a much-altered block of underlying Carboniferous rock. The latter seems more likely. The main vein may be traced for several hundreds of feet. Its width is not constant but swells into chimney-like masses 10 feet or more across. These enlarged bodies also form the main ore shoots. The general trend of the fissuring is N. 10° E., and the dip is 60° to 80° NW. Most of the shoots are composed of massive quartz, but parts are made up mainly of ramifying veinlets and seams cutting the marble. In the quartz masses and quartz stringers are occasional small nests or pockets of galena accompanied by pyrite and tetrahedrite. Much of the early ore produced was largely oxidized and composed mainly of lead carbonate, enriched in secondary silver minerals. Only occasional crusts of malachite can now be found. Work has not extended below the oxidized zone.

Most recent work has been done a short distance south of the old workings. Here a tunnel has penetrated about 15 feet of gouge, 13 feet of manganese and iron-stained rock, and through 30 feet of gossan and leached vein quartz, with occasional scattered granules of galena. The hanging wall has not been encountered, but it is estimated that the lode is at least 50 feet wide.

LAST CHANCE GROUP

The Last Chance group lies in Sec. 22, T. 14 S., R. 24 E., on the high ridge between the head of Grape Creek and Green Creek, about a mile northeast of Cache Peak. The work consists of several cuts and a shallow shaft. The vein is not far from the granite contact and is in the quartzites. Its strike is about N. 40° E. and dip 70° NW. It is a fissure filling and is composed mainly of quartz, in part drusy, with occasional patches of malachite and chrysocolla. The vein is 18 inches wide at the surface. No sulphides have been encountered.

PROPERTIES ON FAIRCHILD CREEK

Much prospecting has been done and is being done on Fairchild Creek on the west slope of the Albion Range east of Oakley, but so far as known no important discoveries have been made. Near the very summit of the range are several pronounced veins with heavy gossan cappings, but assays made by the Bureau analyst fail to show a trace of either gold or silver. These iron cappings are in the schist series of the pre-Cambrian and extend into the Paleozoic limestone that lies not far beneath.

More extensive work has been done lower on the mountainside at the Gray Copper property. Veins on this property do not carry heavy gossans, but consist of a series of stringers of vein quartz with occasional grains of tetrahedrite, mainly altered to malachite and azurite, in Pennsylvanian sandstone (locally quartzite). These quartz seams branch and reintersect, producing a ramifying network generally trending about N. 70° W. and dipping 70° S. Most of the individual seams are narrow, less than two inches, but a few are as much as a foot wide. These show extensive replacement of the shattered or fractured sandstone and the larger usually contains partly replaced fragments of the wall rock. These seams are too widely spaced and the vein material too widely diffused to encourage the likelihood of commercial ore.

A 700-foot tunnel has been driven several hundred feet below the outcrop, but is now caved about 200 feet from the portal. It is driven partly in black graphitic schist and partly in the Paleozoic limestone, and is evidently in the plane of the great overthrust fault. About 300 feet from the portal a layer of the schist is highly pyritic, the sulphide occurring as nodules, seamlets or disseminated, and replacing the schist. Assays reveal that the pyrite carries neither silver nor gold.

PROSPECTS ON BOULDER CREEK AND LAND CREEK

Much prospecting has been done on Land Creek and its tributary, Boulder Creek, about four miles north of Fairchild Creek, also on the west slope of the Albion Range. Pits and cuts are numerous on upper Land Creek, and these have occasionally revealed narrow quartz veins in the pre-Cambrian rocks and the Paleozoic limestone. Some have a little tetrahedrite, usually altered to malachite and azurite.

The principal workings are on lower Boulder Creek, not far above its junction with Land Creek. Field relations are not wholly clear, but the vein appears to represent a replacement of Wells limestone, although the rocks exposed at the surface nearby are pre-Cambrian quartzites. It is probable that the vein is mainly in the block of Pennsylvanian strata from which the pre-Cambrian has been locally eroded or eroded to a very thin shell. The vein trends N. 50° E., but neither its dip nor its thickness could be ascertained. The ore is reported to be argentiferous galena and is associated with white vein quartz. The vein has been explored by a 40-foot shaft from which a 170-foot drift has been driven. These are not now accessible.

MINERALIZATION IN THE BLACK PINE DISTRICT
GEOGRAPHIC FEATURES

Mineralization in the Black Pine district is confined to a comparatively small area on the southeast slope of the Black Pine Range near the head of Black Pine Canyon, along Mineral Gulch, and in several of the gulches east of Black Pine Cone. The entire area covers scarcely more than four square miles, mainly in Secs. 19, 20, 21, 26, 27, 28, 34, 35, in T. 15 S., R. 29 E. The deposits are scattered from the eastern base of the mountain well up toward the main divide, but the most favorable mineralization has been at the very head of Black Pine Canyon and near the base of the Range directly to the east.

Properties at the base of the range may be reached over roads in a poor state of repair. These climb directly up the steep alluvial slopes and are not easily followed because of washouts and sage brush. None of them extend up the main slope of the range, although one is usable for several hundred yards up Mineral Gulch. All connect with secondary roads in the main Black Pine Basin about a mile east of the county line and from there communication may be had with points eastward or with points to the west or south by passing around the ends of the range. It is possible to reach the main property at the head of Black Pine Canyon by automobile with slow, careful driving. This road connects with the Ogden branch of the Old Oregon Trail about four miles from the south base of the range.

This part of the range has no timber other than scattered patches of juniper on its lower southern slope and a few groves of aspen in upper Black Pine Canyon. Necessary timbers in mining must be brought in from the outside or in a roundabout way from the high slopes near the north end of the range.

Lack of water is also a serious handicap to mining and prospecting. All the gulches on the east slope are dry except for a short time after heavy rains, and there are no springs except those several miles from the base of the range in Black Pine Basin. All water needed must be carried several miles from below. Black Pine Canyon is also dry most of the time. Enough water has been trapped in old workings to serve local domestic needs. It is unlikely that enough water could be developed to supply milling requirements throughout the year.

GEOLOGIC FEATURES

The mineral deposits lie almost wholly south of the great transverse fault which crosses the range from Mineral Gulch on the east to Kelsaw Canyon on the west, and are in the block of the Brazer formation on the south side of the fault plane. Folding and faulting in this section of the range have been discussed in the general section on structure, and it is only necessary to restate here the synclinal relation of the strata along Black Pine Canyon and the westerly tilt of the same strata on the east slope of the range as they form the east limb of this syncline. It is necessary only to mention the overthrust character of the block of Brazer limestone and shale as it rests above the underlying block of cherty limestones and sandstones of the Wells formation. Should the veins persist with depth they would probably pass from the Mississippian series into the younger Pennsylvanian rocks below. Mineral deposits have so far been found only in the limestone of the Brazer formation and not in the black carbonaceous shales.

CHARACTER OF THE DEPOSITS

All deposits in the Black Pine district are replacements of limestone, directed along fissures or faults of minor magnitude. These may in general be classed as silver-zinc replacements, zinc replacements, and cinnabar deposits. The silver is contained mainly in tetrahedrite, and for this reason copper might be included in the silver-zinc grouping. Jamesonite also occurs in some abundance in the largest silver-zinc vein and thus the presence of lead might be made the basis for further distinction. The character of the deposits as well as the nature of the metallization is distinctly unlike that in the Stokes district in the Albion Range.

MINERALOGY

GENERAL SUMMARY

The minerals in these deposits are of much greater interest and of much wider range than those in the Stokes district. The primary minerals include quartz, pyrite, sphalerite, tetrahedrite, jamesonite, calcite, barite, cinnabar and realgar. Of these the quartz, sphalerite, tetrahedrite, jamesonite, calcite, and locally barite are the most abundant. The list of secondary minerals is not especially long and includes mainly smithsonite, calamine, malachite, azurite, scorodite, and iron oxides.

PRIMARY MINERALS

Description of the minerals will be taken in order of their deposition in order to stress the paragenetic history of the mineralization and its bearing on the origin or genesis of the deposits.

Quartz: The most abundant gangue mineral of these replacement deposits is white milky granular quartz, usually without trace of chalcedonic or cryptocrystalline structures. Some, however, have local jasperoid facies. Shattering or fracturing of the quartz preceded the introduction of the sulphides. Where the quartz is unfractured by the early movement along the veins it contains no sulphides or other minerals. Its early fracturing always seems necessary before the later metallizing solutions can be introduced and ore shoots delineated. In most deposits the fracturing has been slight and is not readily noticed, but in others the quartz has been reduced to a loose breccia. Later solutions working through the fractured vein matter have deposited sulphides or other minerals in the open spaces, but more generally the later minerals have enlarged the fractures by replacing the quartz on either side or both sides of the original opening. As a result the sulphides form minute reticulating nets in the vein quartz, irregular grains or masses, and lenses and bands. In places much of the earlier quartz filling has been replaced, but in some of the more brecciated zones the pore spaces have not been entirely closed and some of the later sulphides form thin surface grains or coatings. Some quartz has also accompanied the later metallizing solutions, but the quantity is invariably very small and its deposition has persisted throughout the interval of metallic deposition. The later quartz does not show replacement by the other minerals.

Pyrite: Metallization began with the deposition of pyrite. Its usual manner of deposition has been by replacement of the vein quartz and in this process it has developed its own characteristic cubical form. Most of it has been engulfed, however, in the later sulphides, and against them the pyrite has irregular or rounded outlines because of its partial replacement by them.

Pyrite is not notably conspicuous in any of the deposits, although always present. In some its presence is only revealed from microscopic study. Lack of pyrite, except in very small quantity in most of the deposits, may in large part account for the general slight oxidation of the other sulphides, inasmuch as ferric solutions are usually necessary to aid in the attack by atmospheric oxygen. Some deposits do have a heavy iron capping and this feature suggests that some may be heavily pyritized at depth below the present zone of observation.

Sphalerite: Sphalerite is an important ore mineral at the Silver Hills mine and probably occurs in some abundance beneath the oxidized zone at some of the other deposits. Its usual occurrence is as dark brown to pale yellow grains, granules, and irregular masses, also massive seams, showing replacement of the quartz and containing a few scattered remnants of pyrite. Some of the pale yellow grains, however, form distinct tetrahedrons.

Tetrahedrite: The high silver content of some of the ore is probably due to the presence of tetrahedrite. It is one of the main ore minerals at the Silver Hills mine, where it is as abundant as the sphalerite and it occurs as occasional scattered granules in most of the other deposits. In the main occurrence it forms grains, granules, seams, and irregular masses in much the same pattern as the sphalerite. Most of it is the silver-rich variety of tetrahedrite (freibergite) and has a characteristic reddish-brown to nearly cherry-red streak or powder. Its high silver content has been verified by assay. In addition to its high silver, copper, antimony, and sulphur content, it also has appreciable amounts of arsenic. Some of the tetrahedrite forms individual replacement granules or masses in the quartz, but most generally it is directed along the sphalerite masses and in part replaces the sphalerite.

Jamesonite: Jamesonite is most abundant at the Silver Hills mine, but it has been noted in some of the other deposits. It is sufficiently abundant at the Silver Hill mine to give 3.60 per cent lead on average samples of ore. The mineral has a steel-gray color, which readily separates it from the grayish-black tetrahedrite. It tends to occur in compact fibrous masses with individuals in parallel position, as slender needles lining cavities, and as compact and slightly schistose forms. These frequently form aggregates or masses as much as six inches in diameter in some parts of the Silver Hills vein. It is more abundant in some parts of the veins than in others, but it is rarely entirely absent. The mineral is usually intimately associated with sphalerite and tetrahedrite, may locally exceed them in abundance or may be recognized only microscopically within them elsewhere. Jamesonite is another of the minerals that tends toward idiomorphism during replacements and occurs characteristically as laths penetrating quartz, sphalerite, and tetrahedrite, but rarely pyrite. Its replacement of the quartz as idiomorphic laths is especially developed, but its similar relation to the sphalerite and tetrahedrite is only slightly less marked. As the jamesonite increases in quantity the other minerals are partly to wholly replaced and the jamesonite then loses its characteristic form except at margins, although the lath structure may be brought out with proper etch reagents.

Calcite: Some deposition of calcite began before all the jamesonite had been deposited, but most of it came later. Some shattering of the earlier minerals preceded the introduction of the carbonate solutions, for fractures in them are filled with calcite and some of the brecciated sulphides have been completely cemented. A very minor amount of the jamesonite is contemporaneous with the calcite in some of the calcite veinlets, but most of it is clearly older than the calcite. These relations indicate an overlapping in the deposition of the two minerals. The calcite shows irregular distribution in the deposits and forms narrow stringers or veinlets, less commonly seams or masses several feet across. The quantity of calcite in each of the veins is not large when compared with the total filling. Its occurrence is that of a late subordinate gangue mineral.

Several veins composed entirely of calcite occur on the east slope of the mountain below Black Pine Cone. Some of these are 50 feet or more across but contain no sulphides or other minerals. These have mainly formed by direct replacement of the limestone, although it is possible that some have obliterated earlier quartz fillings.

Barite: Barite has been found as a gangue mineral in several of the deposits. In one it forms granular crusts and crystalline aggregates in a very friable vein quartz breccia, and constitutes approximately a fourth of the deposit. In the other it occurs as scattered granular masses and crystals as a part of a massive vein filling with quartz and calcite. Its introduction occurred after the quartz had been fractured and also after most of the calcite had been deposited. Both the quartz and calcite have in part been replaced by the barite. In the first deposit the clefts and pore spaces have not been entirely filled and tabular crystals of barite projecting into cavities are common.

Cinnabar: Cinnabar is present in several of the deposits, but its interest is largely scientific. It is most abundant in the deposits with barite, although its occurrence has been noted in several that have no barite gangue. In the porous and friable quartz-barite chimney it is unevenly distributed and forms thin, sooty or dustlike coatings on quartz and barite crystals or thin films in fractures. In places it is abundant enough to impart a decided reddish color to small pieces of the ore. It also forms occasional small reddish granules and reddish and blackish spots and grains in the massive vein fillings of other deposits. In one of these, the cinnabar also forms a film on tetrahedrite and jamesonite.

Realgar: The association of realgar with cinnabar is suggested both from the color and streak of the cinnabar and from its arsenic content. The color and streak of the cinnabar is in reality intermediate between that of cinnabar and realgar, being neither so scarlet as pure cinnabar nor so orange as pure realgar. In addition the cinnabar gives pronounced tests for arsenic before the blowpipe as well as for mercury and sulphur. The realgar is somewhat subordinate in the admixture of the two minerals.

SECONDARY MINERALS

Smithsonite: Zinc carbonate occurred in sufficient abundance at the Ruth mine to constitute ore, but it is not of value in any of the others. The carbonate has probably been derived from the oxidation of sphalerite and the reaction of the resulting sulphate with the limestone walls. Much of the smithsonite forms crystalline incrustations; some is granular, some reniform, and some is earthy. The crystals are small, generally curved and rough. The color is brown, white, grayish, or tinged with green and blue.

Calamine: Only a very little calamine has been noted in the oxidized ores. It occurs with smithsonite.

Copper carbonates: Malachite and azurite form thin patches or coatings on the oxidized outcrop or for a short distance below. These patches are from the oxidation of the tetrahedrite and are often in contact with partially oxidized residuals of the tetrahedrite. There has been no supergene concentration of secondary copper minerals. Some of the carbonate encrusted residua of tetrahedrite also contain traces of covellite. It appears that covellite is the first product formed as a result of the tetrahedrite oxidation.

Antimony oxides: Several of the pale yellowish or gray antimony oxides remain in the small vugs left on the oxidation of the tetrahedrite. Some occurs on the surface of the partially altered tetrahedrite as well.

Oxidation of the jamesonite has also given a little grayish bindheimite.

Scorodite: Scorodite is an interesting secondary mineral in the upper part of most of the veins, especially those with cinnabar and realgar and to lesser extent in those with tetrahedrite. This hydrous arsenate of ferric iron forms pale leek-green prismatic crystals and amorphous and botryoidal crusts on the surface and in the cracks of the quartz gangue. Its source is from the arsenic in the realgar and in the arsenical tetrahedrite, together with iron from the oxidation of pyrite or other iron minerals.

Iron oxides: Iron gossans are not conspicuous in the Black Pine district and in most veins are lacking. Some veins near the base of the range do have gossan cappings composed of yellow to brownish-black limonite admixed with some reddish hematite and some black pyrolusite. Some parts of the outcrop also contain considerable smithsonite and calamine.

MINERAL PARAGENESIS

Evidence is clear here, too, that crystallization or deposition did not take place simultaneously, but that the several primary minerals were formed successively from solutions which underwent gradual change. Deposition was also sufficiently interrupted at times by movement along the vein to shatter or fracture early minerals and, as the solutions continued to circulate, to deposit new ones, in part by replacing the old. The most abrupt change in the composition of the solutions came near the end when carbonates gave way to sulphates and with the sulphate solutions cinnabar and realgar were brought in.

The essential relationships of the minerals and their sequence of deposition have already been outlined. The order may be given as follows: *quartz*, pyrite, *sphalerite, tetrahedrite, jamesonite, calcite, barite,* cinnabar and realgar. Those italicized are either abundant in all the deposits or predominate in some certain few. Those not italicized are always subordinate or accessory. The chief interruption of the deposition came near the end of the quartz stage, but considerable movement along the vein also occurred during the latter part of the jamesonite deposition and before the introduction of calcite. The sequence given has no unusual features, but is similar to deposits of like kind in other parts of the Cordilleran region. There remains only to call attention to the idiomorphic character of the jamesonite in its replacement of other minerals.

DEPOSITS

STRUCTURAL FEATURES

As the deposits represent replacements of limestone directed along fissures or faults of minor magnitude, they show great irregularity in shape and size. Some deposits are more or less tabular, but most of them swell into occasional chimney-like masses. Most of the veins are less than 12 feet wide and perhaps average about 6, but changes occur so frequently along a given vein that precise figures cannot be given. The main Silver Hills vein at one place is at least 60 feet wide. Such veins can rarely be traced continuously for more than three or four hundred feet along the strike. They may pinch to relatively thin seams, but other chimney-like swellings may appear farther along the fissure. It is very probable that the bodies pinch and swell with the depth as

well as along the strike, for the same vein whose width a short distance below the surface is 60 feet or more has much less width at the surface.

The contact between wall and vein is usually sharp. In most places numerous quartz stringers extend into the limestone wall from the main quartz body and some of the deposits also include large horses of country rock. Replacement of the wall has usually obscured the original fractures or fissures which early directed the solutions.

These bodies are enclosed wholly in limestone so far as present development has shown, but a number of them are near an adjoining bed of shale. In general the shale is not appreciably replaced, but serves mainly to direct the mineralizing solutions.

Mineralization has not been along the major faults in the district but along fractures or fissures of relatively slight displacement. No doubt these fissures are genetically related to the major structural disturbance in the region. Most of them trend in a westerly direction, about parallel to the course of the Kelsaw fault and most likely represent subsidiary fractures developed incidentally to movement along the great transverse fault. These trend from N. 85° W. to N. 55° W., and dip mainly to the northeast. Abundance of finely comminuted and impervious gouge in the major fault planes probably explains the lack of mineralization along them.

Part of the fissuring is perhaps not so much due to movement along the Kelsaw fault as to the bending or plunging of the strata northward against the major fault plane. This movement has developed a distinct crossfold, well brought out in Plate X, A, and part of the openings at least are due to breaking of the strata at the crest of the anticline from tensional forces involved in bending.

There is abundant evidence of recurrent movement along the fissures during mineral deposition and to lesser extent after the major period of ore deposition. Appreciable movement occurred, especially after deposition of most of the quartz and again before the deposition of the calcite. Along some of the veins the movement was sufficiently severe to reduce the early vein filling to a breccia. A little movement has followed the deposition of the ores, but not nearly so much as along the veins in the Stokes district.

VEIN STRUCTURE

The structure of most of the veins is simple and represents massive replacement of the limestone. Such veins are free from banding or crustification, and rarely contain clefts or other openings. The ore minerals are scattered irregularly throughout the vein, although in general they favor some parts more than others, especially such parts as were most broken during the early stages of mineral deposition. The ore occurs mainly as small granules or irregular masses, usually less than an inch, and more or less closely spaced. In some deposits the ore also occurs in small lenses or seams, usually not more than two or three inches wide nor more than a few inches long. As the sulphides have replaced the vein quartz much the same as the quartz replaced the limestone, their occurrence is equally irregular and uncertain. The sulphides are not sufficiently well massed, however, to permit profitable recovery except by milling methods.

Several veins are, however, characterized by intense brecciation, such as is common in many deposits formed near the surface. Fragments have in part been healed with later minerals, but open vugs with drusy surfaces are common. In these the sulphides are confined to the open clefts or fractures, but in the veins sulphide minerals are comparatively rare.

SECONDARY ENRICHMENT

Deposits high on the flanks of the mountains or near the heads of valleys show little evidence of surface alteration or enrichment. Erosion has been in most places more rapid than secondary enrichment processes, and as a result the sulphides appear in the outcrop. In some there may be a slight enrichment very near the surface, but such a zone would be extremely shallow and in most veins of no consequence because the vein was probably too lightly mineralized with primary sulphides at the start. A slight enrichment has probably occurred at the Silver Hills mine, but its effect is not deep and has not greatly increased the tenor of the ore.

Lower on the slopes of the range, oxidation has been more prolonged and secondary enrichment has been a more important factor. Some secondary zinc carbonate ore had accumulated at the surface of the Ruth vein. This enriched ore was bottomed in less than 12 feet below the surface. At that depth, however, sulphides had not made their appearance. Heavy iron gossans at the Hazel Pine suggests the possibility of enrichment, but the zone is apparently shallow, shafts a few dozen feet in depth soon encountering primary vein matter. As a whole, secondary enrichment has been a negligible factor in these deposits.

GENETIC CLASSIFICATION

These deposits have some of the characters of the mesothermal group; that is, of those deposits formed at intermediate temperatures and at moderate depths by ascending thermal waters. Massive replacement of limestone with absence of crustification or banding are especially displayed in the mesothermal group. On the other hand, some of the veins are extensively brecciated and the openings lined with drusy surfaces such as are especially characteristic of the epithermal deposits; that is, of those deposits formed near the surface and at temperatures less than 175° C. or closely approaching those of hot springs. Cinnabar and realgar are especially indicative of epithermal conditions and barite is also more characteristic of the low temperature deposits than in those formed at more intense conditions. Such other sulphides or metallic minerals as are present form under either epithermal or mesothermal conditions. Larsen[1] in discussing one of the cinnabar deposits states that the character of the mineralization—abundant barite, considerable porosity, and the sooty character of some of the cinnabar—indicates a mineralization of no great depth and that the deposit is the result of hot-spring action of comparatively recent geologic age.

There is a possibility that the mineralization has been in two distinct stages, widely separated in age and not even remotely related; the first mesothermal and the last epithermal; and the two associated in the same veins only because later fissuring followed the earlier zones. But there is little evidence of a structural or textural break between the two, for the barite and calcite are nearly contemporaneous and the realgar and cinnabar are not much later. Progressive change in the composition of the solutions with falling temperatures and recurrent opening of the veins seem best to explain the sequence of deposition. Although it is not unlikely that temperatures were near mesothermal when deposition started, through most of the vein-forming period the conditions were probably epithermal and these deposits can be classed in the base-metal epithermal group. Such deposits are not unknown in the Cordilleran region.

AGE AND GENESIS

Genetic relation of the Black Pine deposits to magma is very strongly implied, not alone from the mineral succession or paragenesis, but from the texture and structure of the veins as well. But the absence of intrusive rocks in the district makes it more difficult to trace out such a relationship and also to determine the age of the mineralization. These deposits show little resemblance to the quartz-galena veins in the Albion Range whose genesis has been linked with the intrusion of the late Cretaceous or early Tertiary granodiorite. In fact the char-

---

[1] Larsen, E. S., in Univ. of Idaho School of Mines Bull. 2, Vol. 14, 1919, Tungsten, cinnabar, manganese, molybdenum, and tin deposits of Idaho, by D. C. Livingston, p. 67.

acter of the mineralization is so different in the two localities that the question immediately arises whether they can represent the same epoch of metallization.

Two distinct metallizing epochs have been definitely established in Idaho, one following or accompanying the intrusion of the Idaho batholith and its satellites in the late Cretaceous or early Eocene, and the other following or accompanying the intrusion of granitic rocks in the Miocene or later. Deposits related to the earlier epoch are mainly mesothermal and include the lead deposits in the famous Coeur d'Alene district[1] and in the Wood River district.[2] Deposits related to the later epoch are, on the other hand, mainly epithermal and include most of the gold-silver deposits such as those at Silver City[3] and those in the Salmon River Mountains,[4] as well as the base-metal lead and zinc deposits in the Lava Creek district.[5] Several epochs of metallization are also recognized in adjoining states, including at least four in Nevada.[6] As yet the epochs of metallization have not been fully differentiated in Utah,[7] but the age of the deposits associated with the igneous rocks is said to be essentially that of the igneous rocks to which they are related in origin, and so far as has been definitely shown, the main igneous activity began in post-Cretaceous time. Both extrusion and intrusion continued through middle and late Tertiary well into recent time. Most of the Utah deposits have been assigned to the earliest epoch of igneous activity, but some deposits are associated with lavas of younger age. It seems certain, therefore, that when fully investigated the Utah metallization will be related to an early Tertiary epoch when the copper and lead-zinc veins and replacement deposits of mesothermal characteristics were formed, and also to a later Tertiary epoch when veins with epithermal affinities were deposited in Tertiary extrusive rocks. Both of these epochs are represented in Nevada, and in addition there is one at the close of the Jurassic and another in the late Tertiary. These several epochs of mineralization in the surrounding region provide ample grounds for believing that perhaps two of them are present in Cassia County, one in the late Cretaceous or early Tertiary as already discussed, and one near the mid-Tertiary.

[1] Ransome, F. L., and Calkins, F. C., Geology and ore deposits of the Coeur d'Alene mining district Idaho: U. S. Geol. Survey Prof. Paper 62, 1908.

[2] Umpleby, J. B., Westgate, L. G., and Ross, C. P., Geology and ore deposits of the Wood River region, Idaho: U.S. Geol. Survey Bull. 814, 1930.

[3] Piper, A. M., and Laney, F. B., Geology and metalliferous resources of the region about Silver City, Idaho: Idaho Bureau of Mines and Geology, Bull. 11, 1926.

[4] Ross, C. P., Ore deposits in Tertiary lava in the Salmon River Mountains, Idaho: Idaho Bureau of Mines and Geology, Pamphlet 25, 1927.

[5] Anderson, A. L., Geology and ore deposits of the Lava Creek district, Idaho: Idaho Bureau of Mines and Geology Pamphlet 32, 1929.

[6] Ferguson, H. G., The mining districts of Nevada: Econ. Geol., Vol. XXIV, 1929, pp. 115-148.

[7] Butler, B. S., Loughlin, G. F., and Heikes, V. C., The ore deposits of Utah: U.S. Geol. Survey Prof. Paper III, 1920, pp. 95, 100.

In most places the mid-Tertiary deposits are in association with volcanic rocks. Their absence in the Black Pine district makes correlation much less direct, yet, as has already been stated, the Tertiary strata and lavas at one time blanketed the region and they have been subsequently eroded from some of the higher arched parts of the district. This can account for their present absence from the mineralized area.

It therefore seems most likely that the mineralization in the Black Pine district belongs to the younger of the two main epochs of metallization now recognized in Idaho or to the Tertiary epoch. The close grouping of the deposits in the Black Pine Range suggests the occurrence of an intrusive body some distance below the surface as the control and source of the mineralizing solutions. As the volatile constituents left the magma they found their way upward through the fractured rock, especially along the minor fractures or fissures in the broad zone developed along the major transverse fault. Deposition of minerals apparently did not occur until the temperatures of the solutions had been much lowered or were not far from that of thermal springs. As these solutions encountered the limestone walls they reacted chemically and replaced the limestone by quartz. Movement along the fissures concurrent with deposition maintained open channel-ways for the passage of these solutions. Deposition from the solution was progressive. Earlier minerals formed were in part unstable, as the composition of the solutions changed, and these were partly replaced by succeeding minerals. These solutions were at first mainly siliceous, then became enriched in carbonate, and finally enriched in sulphate. Cinnabar and realgar were the final products of the metallizing solutions.

COMPARISON WITH DEPOSITS IN OTHER REGIONS

The Black Pine mineralization has some features in common with the mineralization described by the writer in the Lava Creek district.[1] In the latter, however, the deposits occur as replacements of volcanic rock, mainly andesitic lavas, but a few occur as replacements in the underlying Paleozoic limestones. These deposits are also of the epithermal base-metal type and the chief values are in lead, zinc, and silver. The veins are grouped near bodies of intrusive Miocene (?) granite. Deposits in both areas probably belong to the rather exceptional base-metal epithermal type listed by Lindgren,[2] in which chalcopyrite, galena, zinc blende and tetrahedrite are associated in an abundant gangue of quartz, carbonate, or barite, but with principal values

---

[1] Pamphlet 32, op. cit.
[2] Lindgren, Waldemar, Mineral deposits: McGraw-Hill Book Co., 1928, pp. 526; 580-592.

usually in gold and silver. Such deposits are especially well represented in the most interesting metallogenic province of the San Juan region in Colorado, including the mining districts of Telluride, Ouray, Silverton, Lake City, Rico, Needle Mountains, La Plata, and Creede. Some are rich in galena, sphalerite, and tetrahedrite, but the yield is principally in gold and silver. Deposits of similar kind in Utah are also associated with Tertiary extrusives and consist of similar base-metals in a quartz-calcite-barite gangue. In both the Mercur and the Marysvale districts, the base metals and the quartz-calcite-barite gangue are accompanied by some cinnabar and realgar.

OUTLOOK FOR THE DISTRICT

The epithermal deposits constitute the source of a large part of the world's production of gold and silver and they contain the spectacular bonanzas of the Cordilleran region, of which examples are found at Cripple Creek, Goldfield, Virginia City, Silver City, and many other districts. Base-metals are present plentifully enough in places, as in the San Juan region, but the mines are rarely worked for these alone. Silver and gold are the profitable metals and the others are usually only by-products or are not recovered at all.

The Black Pine district gives little promise of ever becoming a producer of base metals alone. It is true that some zinc has already been produced, but the zinc ores were from the enriched surficial zone and the supply has been largely exhausted. In all deposits, except one, the zinc blende is only an accessory mineral. The same may be said of the lead mineral, jamesonite, and as this mineral is not so rich in lead as is galena, it is probably not in sufficient abundance to warrant mining for its lead alone. The future of the district must depend mainly on the precious metals in the ores, especially silver. The tetrahedrite is sufficiently rich in silver and is sufficiently well distributed in the Silver Hills vein to offer attractive inducements, especially if silver should command a better price than it does at the present time. In addition to the silver at the Silver Hills mine, the lead and zinc are also especially worthy of recovery. So far, only the Silver Hills mine has disclosed sufficient metallization to make the district attractive to further development.

PROPERTIES

SILVER HILLS MINE

The Silver Hills mine lies very near the crest of the ridge between Black Pine Canyon and East Dry Fork on the Black Pine Canyon side, in Sec. 26, T. 15 S., R. 29 E. The veins outcrop almost on the divide, but the tunnels and camp are several hundred feet below at an elevation near 7,000 feet A.T. This property is connected by road with the southern branch of the Old Oregon Trail. The connecting link follows the bottom of the canyon from the edge of the mountain to within a short distance of the camp where it must ascend a comparatively steep short switchback. The upper part of the road, though steep, offers no greater hindrance to travel by automobile than the numerous crossings of the dry creek bed lower in the canyon. This property is the main one controlled by the Silver Hills Mining Company whose organization dates from March 31, 1920. The Silver Hills mine was located about 1880 and some ore shipped about 1894. Neither the amount of ore produced at that time not subsequent production, if any, was learned. This mine has the largest and most highly mineralized vein in the district, and has been much more fully developed than any of the others. Development consists of more than 3,000 feet of underground workings, in addition to a long crosscut only recently started. In addition to this mine, the company has acquired many surrounding claims, including most of those in Mineral Gulch on the east.

The Silver Hills mine affords the most ideal example of the Black Pine type of mineralization. It is strictly a replacement of limestone and in consequence has the usual irregular shape and form characteristic of deposits of such kind. Its position is directed along a line of fissuring about parallel to the Kelsaw fault, which lies about a half a mile or more to the north, and it is in the massive light gray limestone so widely distributed in the southern end of the Black Pine Range. The fissuring here is apparently not so much related to movement along the Kelsaw fault as to the bending or plunging of the massive limestone beds against the fault plane. This downward plunging of the normally northerly trending strata amounts essentially to a large cross fold or arch, and it is in fractures near the axis of the arch that the most extensive mineralization has occurred. In the vicinity of the deposits the limestones trend about N. 75° W. and dip about 16° N. Some of the veins and minor quartz stringers strike in about the same direction as the limestones, but the main vein strikes about N. 85° W. and dips 70° N. Beneath the limestone occurs a thick series of calcareous and carbonaceous shales into which the veins must eventually pass with depth. These shales appear in each of the tunnels on the property. In the principal underground disclosure they practically form the footwall of the vein.

Replacement along the fissures and fractures has been mainly by quartz, but in the main deposit the quartz is accompanied by some late carbonate and by grains, lenses, or small irregular masses, and occasionally wider compact seams, of ore minerals, notably sphalerite, tetrahedrite, and jamesonite, and to lesser extent pyrite. Assay made of more or less average ore from the upper tunnel shows the following values: gold 0.03 ounces, silver 42.4 ounces, lead 3.6 per cent, copper 1.2 per cent, zinc 8.4 per cent, and antimony 4.5 per cent. Assay of selected high-grade will give returns of several hundred ounces of silver per ton. Most of the deposit consists of massive vein material, together with some stringers of quartz and also larger blocks or horses of limestone. The vein is perhaps 30 or 40 feet wide at the surface, but its true width in the workings 75 feet below the outcrop has not been determined, although it probably exceeds 60 feet. At the surface the vein may be traced for several hundreds of feet, but it shows much pinching and swelling and no doubt maintains the same character with increasing depth, at least until shale is encountered, when pinching may be very likely. The cropping contains scattered or occasional grains or granules of tetrahedrite, less commonly sphalerite, and also occasional patches of malachite and azurite. There is very little sign of oxidation or enrichment on the level 50 feet below the cropping although some vugs show, but at 75 feet the ore is wholly primary.

The vein may be examined underground only in the upper tunnel, whose vertical distance below the outcrop is about 75 feet. This tunnel passes through about 90 feet of shales; black carbonaceous shale near the portal and lighter colored calcareous shale farther in; before encountering a zone of more massive limestone with barren quartz stringers. At a point about 114 feet from the portal some of the stringers contain a little of the ore minerals, but it is not until after passing through about 50 feet of alternating shales and limestones (also with quartz stringers) that the main ore body is cut. From this point a short drift has been run a few feet to the southwest on a well mineralized zone or streak of high-grade; also a winze of undetermined depth has been sunk, and drifting has been carried to the northeast. Drifts to the northeast branch and pass diagonally along the vein, but none have extended to the walls. As much as 60 feet of the vein have been exposed, all in ore. Some of it includes masses of country rock and stringers in limestone, but the larger part is massive vein quartz and calcite with a scattering of ore minerals, sphalerite usually predominating, closely followed by tetrahedrite and subordinately by jamesonite. There are occasionally some high-grade seams and more closely grouped masses of the ore-minerals. No more exploratory work has been done along the vein than across it and the true limits of the shoot cannot be fixed. Most drift faces end in ore or mineralized vein matter. From one of the drifts a raise connects with the outcrop at the surface. From another drift a raise connects with a level 12.5 feet above. This level consists of a single drift a little over 100 feet in length and driven about N. 55°W. It cuts much vein matter and numerous stringers in limestone. Some parts are well mineralized. A raise also connects with a still higher level 14 feet above. This one is parallel to the one below, but is only about half as long. Stringers of quartz are also numerous on this level, some of them with metallic minerals and some of them with vugs from which the ores have been leached.

A second tunnel has been driven about 100 feet below. It is now blocked to serve for water storage. This tunnel also discloses the shale series beneath the vein. It crosses the vein, but no drifting has been done on either side. It is unfortunate that this tunnel was not available for study, as it should give valuable clues as to the character and persistency of the minerals with depth, also the effect of the shale series on the size of the ore body. A long crosscut had been started about 550 feet below the level of the upper tunnel and had penetrated several hundred feet of black carbonaceous shale at time of the writer's visit. No attempt was made to measure the distance the tunnel must be driven before the vein chould be encountered, but it will probably have to be extended two thousand feet or more.

The upper workings are not sufficiently adequate properly to disclose the ore body. Conditions shown there amply justify further development, especially to determine the width and length of the mineralized body. Much more work could be expended in the tunnel 100 feet below to determine what effect the shale might have on the continuance of the ore shoot or on its size and position. The ore shoot is certain to be irregular and erratic, as is indicated by the irregularity of the vein on the surface and as suggested in the several levels of the upper workings. Other veins showing at the surface are probably in part those encountered as stringers and in wider bodies in the workings underground. No doubt much of the contiguous country rock will contain irregular veins, stringers, or chimneys of quartz, more or less thoroughly metallized.

RUTH MINE

The Ruth Mine, acquired by the Silver Hills Mining Company in 1929, lies not far above the east base of the Black Pine Range at an elevation near 6,500 feet, in Sec. 26 or 27, T. 15 S., R. 29 E. Its position is on the crest of a secondary ridge between Mineral Gulch and the first large gully on the south that opens to the main Black Pine basin. It is reported that six cars of rich zinc carbonate ore were mined on the property during the World War. This ore was mined directly from the outcrop. There has been little work done underground.

This deposit occurs as a replacement of massive light gray limestone, such as outcrops at the Silver Hills mine. The original mineralization followed a narrow zone of fissuring whose trend is N. 60° W. and dip about 40° N., cutting the general trend of the limestone, whose strike varies from N. 80° W. to N. 80° E. The fissuring has extended across the ridge, well down into Mineral Gulch and has been followed by cuts for 110 feet. Along this zone the vein swells into small chimney-like masses or lenses, sometimes measuring up to 12 feet across, but usually less than 4 feet. In places it pinches out completely. Other cuts and short tunnels on iron-strained croppings farther to the north-west suggest that the vein continues for a long distance in that direction.

The main ore is smithsonite and it occurs mainly as crystalline incrustations or as botryoidal crusts, as a filling in former openings, or as a massive granular replacement of the limestone walls. Its color is ordinarily white or grayish, but some of it is tinged with green and blue. Some ore still remains and some of the lumps on the surface are several feet across. In some places there are also patches of malachite and azurite, whose presence suggests that tetrahedrite may be among the primary ores. None of the cuts have actually extended into the primary zone, but some tiny grains of brownish sphalerite were noted in some of the material on the dump, likewise an occasional grain of cinnabar and realgar. Other vein matter noted consists of calamine, a little scorodite, and a little vein quartz. Oxidized ore has been mined on the surface for nearly 100 feet.

HAZEL PINE PROPERTY

The Hazel Pine property lies at the base of the Black Pine Range in Sec. 26, T. 15 S., R. 29 E., about one mile south of the mouth of Mineral Gulch. It is reported that 14 cars of ore have been shipped from the property, but these figures could not be further verified nor could the character of the ore shipped be ascertained. The ore is said to contain gold, silver, and lead.

This vein, like the others in the district, is a replacement of limestone and exhibits the usual great irregularity in size and shape, frequently swelling into chimney-like masses. Its course is directed by fissuring or brecciation in the limestone and its strike is about N. 55° E. and its dip steeply northwest. Much of the replacement is by white granular quartz, but with the quartz is considerable vein calcite. No primary ore minerals were seen, but the ore is entirely oxidized and consists of yellowish and reddish iron oxides, black manganese oxide, with occasional streaks or crusts of smithsonite and calamine, and less commonly scorodite. Much of the cropping can be regarded as a gossan and is unusually heavy in iron oxides. Some of the iron has nearly a vermillion or crimson color. Assays were made on a number of samples collected from various parts of the cropping and from ore on the dump, and although some showed values in zinc and a trace of lead, none contained gold nor more than an ounce of silver.

A notable amount of work has been done near the south end of the property. Here the vein has been prospected by a series of shafts, tunnels, drifts, and glory holes, most of which are now inaccessible. One of the shafts descends at least 50 feet. Some of the tunnels open into large chambers a dozen feet or more in diameter, and these no doubt give clue to the shape and size of the ore shoots. Apparently much vein matter has been removed from the workings.

More work has been done about 100 yards to the northeast where the vein has been cut by a small gully. This has afforded better means for working the deposit and several tunnels have been run into the vein from the side of the gully. These short tunnels also open up into chambers 12 to 15 feet in diameter and expose much iron oxide and numerous quartz veinlets in limestone.

VALENTINE PROPERTY

The claims of the Valentine Cinnabar Company also lie near the east base of the Black Pine Range at an elevation of 6,000 feet A.T. in Sec. 35, T. 15 S., R. 29 E. This property has been described by Larsen,[1] who states that the cinnabar deposit is at the top of a limestone bed beneath a shale horizon and has been formed through a replacement of the limestone. The deposit has the form of a chimney and dips with the bedding of the sediment at about 30° to the west. In horizontal section it is elliptical or lens-like, is about 10 feet long and 5 feet wide, with some stringers of mineralization extending along the bedding for 10 feet or more in both directions beyond the main chimney. It has been opened up for a depth, measured along the dip, of about 40 feet.

The material of the mineralized chimney is porous and friable, breaking up into coarse sand, and made up in large part of barite and quartz. Tabular crystals of barite commonly project into the

---

[1] University of Idaho School of Mines Bull. 2, Vol. 14, 1919, pp. 65-67.

cavities. The cinnabar is unevenly distributed and forms a very thin, sooty or dust-like coating on the quartz and barite crystals of the cavities. In places it is abundant enough to impart a decided red color to the rock, but nowhere forms more than a very small part of the vein material. This cinnabar also carries realgar. Near the surface there is much scorodite. Some gold is reported to be present in the deposit, notably in some of the narrow seams which extend along the bedding from the main chimney. A sample of quartz containing some specks of cinnabar and realgar gave on assay 0.05 ounces in gold and 2.7 ounces in silver per ton. A prospect hole about 50 feet to the east of the chimney has been sunk to considerable depth and has exposed some crushed, non-ironstained material that is said to carry gold and some silver and lead.

MILLER PROPERTY

The Miller property lies high up the slope of the range at an elevation above 7,500 feet A.T., a short distance below the top of Black Pine Cone, in Sec. 34, T. 15 S., R. 29 E. This deposit is also a replacement of limestone along a fissure or breccia zone whose strike is N. 80° W. The vein ranges from two to six feet in width, with stringers extending to greater distances. It may be traced for several hundred feet by the line of cuts and short tunnels. Mineralization is mainly by quartz but includes a little tetrahedrite, jamesonite, cinnabar and realgar, and some thin patches of their alteration products. Barite also occurs in the ores. Workings consist essentially of two short tunnels, one about 50 feet long and the other about 60 feet. These have a vertical range of 50 feet.

PROSPECTS ALONG MINERAL GULCH

Several veins have been explored along Mineral Gulch, but no work has been done on them for years. They are now controlled by the Silver Hills Mining Company. These show little difference from those already described, but occur as replacement of limestone, mainly on the south side of the Kelsaw fault. Mineralization has been chiefly with quartz, and to lesser extent with calcite and barite. In addition, some contain scattered granules of tetrahedrite and patches of secondary copper carbonates, and, occasionally, crusts of zinc carbonate. Most of the tetrahedrite is the freibergite variety and selected speciments of nearly pure mineral on assay give very high values in silver. Some of the veins are reported to have values in gold. These deposits are scattered from the mouth to the head of the canyon.

One deposit lies directly across the canyon from the Ruth about 100 feet above the canyon floor. This one has much quartz, some calcite, and scattered granules of tetrahedrite. It is also reported to carry gold. Another lies a half mile or more up the canyon and consists of a large quartz-calcite vein with scattered patches of azurite and malachite and some thin crusts of smithsonite. This vein is about 12 feet wide where exposed and strikes N. 70° W. and dips 40° W. Others are nearby but have little to show. Several lie very near the head of the canyon. One of these consists of a heavily iron-strained gossan encrusting a gangue of quartz, calcite, and barite.

OTHER PROSPECTS

Mineralization is also shown near the head of East Dry Fork and along the ridge between the Silver Hills mine and War Eagle Peak. The most striking occurrence is on a ridge between two tributary gulches near the head of East Dry Fork in Sec. 21, T. 15 S., R. 29 E. This occurrence consists of an immense iron capping or gossan zone about 200 feet wide and 200 yards long, striking about N. 55° W., and occurs partly in sandstone and partly in limestone. No openings have been made on the iron capping and little could be learned of its character other than that it consists mainly of limonite in the form of black limonite varnish and shows few vugs and little porosity. The presence of so much iron oxide should excite curiosity as to the nature of the primary mineralization beneath.

Another prospect lies near the top of War Eagle Peak at an elevation above 8,200 feet A.T. At this place a remarkably persistent replacement vein in limestone may be traced in a westerly direction for a long distance. The vein is more than 40 feet wide in places and has been explored by glory hole and a 50-foot shaft. The vein material consists essentially of white granular quartz with very scattered patches of azurite and malachite and occasional grains of partly oxidized tetrahedrite.

Several disclosures have also been made on the north side of War Eagle Peak and several to the east near the divide between East Dry Fork and Black Pine Canyon. These show quartz replacing limestone and traces of secondary copper minerals.

BUILDING STONES

GENERAL FEATURES

The district is supplied with an abundance and a large variety of rocks suitable for building materials. Some of these have been utilized in the past, especially when transportation facilities were such that it was cheaper to quarry and use nearby rocks than to bring in lumber from distant sources. Although in more recent years bricks and concrete have found greater favor as structural materials, nevertheless various kinds of rocks continue to be quarried and constitute a large economic resource. Some of the materials have very pleasing qualities, such as are desired in residences, churches, banks and public buildings. Such materials have been used in schools, churches, and business buildings in Oakley, and have also found favor in Albion in business buildings and in the construction of the Albion State Normal School. Rock has also been used elsewhere to more limited extent, particularly in the school building at Malta and in several buildings in Declo. These places are all near easily available sources. Little rock has been used at Burley, however, partly because it is about 15 miles from the available material, but mainly because structural materials are more readily and perhaps more cheaply obtained from a local brick plant. Nevertheless there will always be some local demand for rock as building material, and such demand may increase, especially among home-builders and for municipal structures.

Almost all available kinds of rocks are used as building materials in this region, but most demand has been for the more easily quarried and trimmed Tertiary tuffs and lavas. Quartzite has been used in one instance; Snake River basalt has been employed at Declo, especially for foundations and back walls; marble has been used to very limited extent. There has been no attempt made to utilize the Carboniferous sandstones and limestones or the granitic rocks. In general but a single kind of rock has been used in each construction, but in some few buildings very pleasing results have been obtained by combining one kind of rock with another. With the wide range in color and texture afforded in these materials, there is chance for very artistic and striking combinations, such as are especially desired in residences. The character of and quality possessed by each of the materials will be discussed in detail.

VOLCANIC TUFF

Tuffaceous beds in the Tertiary series have been more extensively used for building purposes than any other variety of rock, mainly because of the ease with which the tuff can be quarried and trimmed into blocks. This kind of rock is widely distributed in the County, as discussed in the section of the report on the general geology of the district, but is mainly concealed beneath the latite flows. Exposures, however, occur at regular intervals near the base of fault scarps, in the base of landslide blocks, and in the gulches or canyons cut through the capping lava. In a few places it presents white bold outcrops, but in most places the tuff is beneath a mantle of hillwash of variable thickness. Quarries have been opened in only a few places and have supplied material used in buildings at Albion, Oakley, and Malta.

At first glance the tuff appears wholly inadequate as a building stone, for the material is only slightly compacted or indurated, and is not well cemented, if at all. It appears to be somewhat crumbly and small fragments may be broken between the fingers. Yet it seems to harden slightly after seasoning and has been used in the walls of two-story buildings (Plate XVI, B). Most of the tuff shows bedding (Plate XVI, A), though not conspicuously laminated, but, fortunately, parts with some difficulty along most of the planes. It is cut by few widely spaced fractures or joints approximately at right angles to the bedding. These fractures lack regularity and therefore cannot be utilized to advantage in systematic quarrying. Much of the tuff can be pried loose from the ledge without the use of explosives and is easily trimmed into rectangular blocks of appropriate sizes.

The tuff has a pleasing white to gray color and from casual inspection resembles very light gray sandstone. The white is not as greatly desired, however, as is an intercalated stratum of pale greenish yellow color, whose desired quality is in its more resistant character rather than in its color. The white or light gray tuff has the composition and texture of a fine to coarse grained pumiceous ash and is composed almost wholly of colorless shreds of glass, occasional feldspar fragments, and a few small specks of black mica. These are only loosely cemented and the rock has a high porosity, the individual shreds of glass being readily detached by rubbing. Much of it is plainly laminated and the individual layers are usually an inch or two thick, although some are 18 to 36 inches.

The greenish yellow-tinted tuff differs considerably in composition as well as in appearance from the white or light gray tuff. Its texture is more granular and its fragmentary glassy constituents are not nearly so well observed. Fragments and crystals of feldspar are plainly visible. Under the microscope it shows more than half feldspar fragments (oligoclase-andesine), some augite fragments, and a few magnetite grains embedded in a glassy and in part clayey matrix. The glassy matrix is full of large gas cavities which add greatly to the porosity of the rock. Chlorite is also present in minor amount and an occasional grain of zircon. Some of the clay seems to have developed from alteration of the augite, but much of it was perhaps contributed at the time of deposition of the tuffaceous material. Most of the feldspar is surprisingly fresh. It would weather rapidly perhaps when wet, but in so dry a country as this none of the materials constituting the rock detract from its adaptability as a building stone, and actual use in the region shows that the stone does not weather seriously under local climatic conditions. Enough clay is present to impart a distinct clayey odor to the tuff. It has been demonstrated that rocks containing noticeable quantities of clay minerals are not well suited for many structural uses,[1] as the effect of these minerals is to induce a gradual swelling and spalling of the rock when moistened, with consequent weakening. This is not a serious factor in the utilization of the tuff in Cassia County, because of the aridity of the region. Unfortunately, such coarse, granular, better compacted layers are not numerous in the series.

None of the tuff was submitted to physical tests, but the results, if such tests had been made, would probably have been similar to those obtained in the building stone (tuff) from Challis,[2] Idaho. Arithmetical averages of several tests from the Challis deposits are given below. The rock there tested is greatly similar to the granular pale greenish yellow tuff in Cassia County.

RESULTS OF PHYSICAL TESTS OF BUILDING STONE (TUFF) FROM CHALLIS, IDAHO

Compressive Stength:
Dry, perpendicular to bedding (3 tests), pounds per sq. inch..............10,880
Dry, parallel to bedding (3 tests), pounds per sq. inch...................11,446
Wet, perpendicular to bedding (3 tests), pounds per sq. inch............. 4,245
Wet, parallel to bedding (3 tests), pounds per sq. inch.................. 4,136

Absorption by weight (12 tests), per cent 13.70
Apparent specific gravity (6 tests)................................. 1,742
Weight (dry, per cubic foot), pounds................................ 109

The Challis rock was subjected to weathering tests. Four out of six specimens showed extensive disintegration upon 123 freezings, and the remaining two withstood 162 freezings with no recognizable change. All showed complete disintegration after having been frozen 255 times. The testing bureau concluded that the tuff is more resistant to weathering than some of the poorer grades of limestone now on the market.

Comparative data on the strength of tuffs are scarce. The compressive strength of the tuff from Challis is about that of similar material from Lilliwaup, Washington,[3] but the absorption ratio is

1 Loughlin, G. F., Usefulness of petrology in the selection of limestone: Rock Products, Vol. 31, 1928 pp. 52-53.
2 Behre, C. H., Jr., Tertiary volcanic tuffs and sandstones in the upper Salmon River Valley, Idaho: U.S. Geol. Survey Bull. 811-E, 1929, pp. 244-245.
3 Shedd, Solon, The building and monumental stones of Washington: Washington Geol. Survey Ann. Report for 1902, Vol. 2, 1903, pp. 49-51.

greater. Merrill[1] mentions a tuff in California which, though having a greater density, possesses a dry crushing strength of only 7,469 pounds to the square inch at right angles to the bedding. Compared to several building stones, the Challis tuff has a compressive strength when dry as great as many limestones and sandstones, despite its low density. However, the limestones and sandstones exhibit no such noticeable decline in strength when wet; this reduction in strength probably being due to the presence of the clay mineral.

The white pumiceous tuff in Cassia County, however, cannot be compared with the Challis tuff, as it is of wholly different character and texture. Very likely its compressive strength is much less, both when dry and wet, and its absorptive capacity is probably greater. Nevertheless, its strength appears to be in excess of local requirements.

Both varieties of tuff have served well in the bank building in Albion (Plate XVI, B), in one of the stores, and also in several of the local residences. These structures have been built for many years, and the rock shows little sign of weathering or disintegration. The rock was quarried at a place about seven miles southeast of Albion in Sec. 35, T. 12 S., R. 25 E., about one-half mile from the Elba road.

The quarry is near the west base of the Malta Range and appears in a gully beneath a thick capping of black glassy columnar latite, which grades upward into the platy grayish porphyritic facies. The quarry face has been extended along the tuffaceous beds for a little more than 500 feet. Beyond, the tuff is concealed by the talus from the latite flow above. The tuff and overlying lava dip about 5° westward and a short distance down the slope disappear beneath the alluvium between the Malta and Albion ranges. The lower part of the stratified series is concealed by landslide and hillside debris, but a hundred feet of beds may be safely estimated at the quarry site. Only the upper 20 to 25 feet of it is easily available and utilized as building stone.

The tuff is directly overlain by the porphyritic glassy lava and almost seems to merge with it. The upper 15 feet is not used because of its more or less softened condition due to weathering. This has been stripped back. The lava above because of its columnar jointing, has receded more rapidly than the strata beneath, and in so doing has provided a quarry depth of some size. As the quarry face extends into the hill, the overburden, and eventually the lava capping, will assume such proportions that the tuff can no longer be easily or profitably quarried. The amount still easily available can supply local needs for many years.

[1] Merrill, G. P., Stones for building and decoration: New York, 1910, p. 507.

A section of the tuffaceous series as exposed in the quarry face (Plate XVI, A) is given below:

SECTION IN QUARRY AT WEST BASE OF MALTA RANGE
SEVEN MILES SOUTHEAST OF ALBION

	Feet
Latite flow, black glassy base, columnar structure, grading upward into grayish platy rock	100.0 +
Tuff, admixed with fragments of latite	1.0
Tuff, grayish, white, pale buff, fine to coarse, glassy pumiceous fragments, in part sandy in layers 1 to 20 inches thick, in part finely laminated; lies above present quarry face	11.0
Tuff, coarse to fine, mainly medium, grayish, pumiceous, in layers five to twelve inches thick; partly quarried	6.0
Tuff, medium grained, pumiceous, with two irregular seams of coarser glass; utilized	2.5
Tuff, alternating layers two to six inches thick of fine to coarse grain, grayish, pumiceous, yielding thin blocks	6.0
Tuff, granular, pale greenish yellow in layers one to five inches thick, alternating with grayish pumiceous layers of equal or greater thickness	2.5
Tuff, coarse granular, greenish yellow, feldspathic	1.5
Tuff, medium grained, grayish, pumiceous	1.0
Tuff, coarse granular, greenish yellow, feldspathic, with grayish pumiceous layers	1.5
Tuff, pumiceous, from quarry floor to base of gully 50 feet lower	50.0
Total	185.0

The tuff remains firm in the quarry face and maintains vertical walls. All parts may be readily quarried. The blocks are usually plainly dressed, but are adapted for taking a chisel-dress. As the more desirable granular feldspathic tuff forms such a small proportion of the whole series, it is used mainly as a trimming or in cornerstones and the pumiceous facies is employed in the back walls. This adds a pleasing color contrast in the buildings in which it is used.

Similar tuff has been quarried a few miles southwest of Oakley, but the quarry is now beneath the storage waters impounded in the Goose Creek reservoir. It is wholly the pumiceous variety. One of the garages in Oakley has back walls of this tuff. The front is finished in brick. Tuff is also used in cornerstones and for trimming in the Latter Day Saints' Church in Oakley (Plate XVII, B), and in the construction of several houses and business buildings. Such material is still easily available in the canyons south and west of Oakley.

The quarry from which the tuff used in the school building at Malta was obtained was not visited. This tuff presumably was procured from the base of the fault scarp or from a landslide block at the foot of the Malta Range. Building stone of such character may be had in most of the region of volcanic rocks with but little search or prospecting beneath the mantle of hill wash.

The precipitation in Cassia County is too low to be damaging to the stone used above the capillary reach of water from the ground if the stone has been properly seasoned before use, and especially if it is well protected from direct rain by an overhanging roof. Hence the tuff, despite its lightness and porosity, wears well under local weather conditions. Furthermore, the capacity of the porous rock to resist freezing and weathering is in part indicated by the tendency toward complete saturation. The tuffs, despite their high porosity, show a markedly low saturation, which lends resistance to disintegration by freezing.

Tests on the Challis rocks show that the absolute compressive strength of the tuff—a minimum of 4,000 pounds to the square inch, even when wet—is far in excess of local requirements. The strength of a stone under water depends upon its constituents and texture. A rough measure of its suitability is given by the ratio of its compressive strength when water saturated to its strength when dry. For the tuff from Challis, tested by the Bureau of Standards, this ratio is 0.37—a low figure. For such uses as bridge piers or even house foundations, in a water-soaked subsoil, the tuffs near Challis would not be satisfactory, and the same conclusion may be drawn with the tuffs in Cassia County. Hence special precautions should be taken if use in foundation is contemplated; spalling or disintegration may be avoided if the stone is coated with cement mortar. At heights of two feet or more above the ground there should be no danger of spalling or disintegration. Such action is much less likely in the pumiceous tuff than in the coarsely feldspathic with its admixture of clay.

QUARTZ LATITE

The quartz latites associated with the Tertiary tuffs have proven most satisfactory as a building stone. The rock is not as easily quarried or trimmed as the tuffs, but when once quarried and shaped it has a desirability and strength equal to the best of building stones and is suited for most purposes where resistance to crushing strength is especially desired. In addition, it has a pleasing color, normally pinkish or light purplish-gray, which adds much to its attractiveness. Most of that quarried has only a slight porous structure, which does not detract from its value for use in the walls of buildings or as a foundation stone. None of that used shows signs of weathering.

Quartz latite is widely distributed in Cassia County, as discussed in another section, but not all of it is suited as a building stone. The lower part of the flows is generally vesicular, black, glassy, and although some of it has been quarried and used locally in foundations, its porosity and brittleness, together with its unattractive color, do not make it desirable. Above, the flow changes gradually from black porphyritic glass to pale pinkish or reddish aphanitic rock, and in some of the thicker flows into a gray fine-grained crystalline rock, in all parts porphyritic. Generally the pinkish tinted rock is quarried, but the grayish also possesses pleasing qualities. Unfortunately, so much of the flow, including that with attractive color and suitable texture, has a thin platy parting. This structure makes it impossible to procure blocks of desired size except in occasional massive bands or layers some distance below the surface of the flow. Where erosion has removed the thin platy part of the flow the more massive parts not far above the black glassy zone may be quarried. In an earlier discussion of the petrography of the latites it was pointed out that the pinkish latites have mainly a glassy groundmass, but with a better developed crystallinity than in the underlying black obsidian facies; whereas, the grayish rock is nearly entirely crystalline. The platy character appears to increase in most places with the crystallinity of the rock, and for this reason the pinkish facies with its lesser tendency to develop parting planes has been most generally used as a building stone. Some of the thinner flows are perhaps better suited for exploitation than the thicker ones because of the proportionately greater quantity of pinkish or purplish tinted rock.

This lava has been quarried and used as a foundation stone for most of the buildings on the campus of the Albion State Normal School. As such it forms the first two to eight feet of the wall above the ground, and in the gymnasium building nearly the entire lower story. Because of its relatively low porosity it can be used as a foundation stone where capillary action would rapidly destroy tuff if used for the same purpose. Because of the difficulty and expense of quarrying and trimming, it is not used in the superstructures of the building. Red brick is used instead. Nevertheless, with its minor role it adds to the attractiveness of the buildings. This rock is quarried near the edge of the campus on the top of the hill immediately north of the town of Albion, only about 200 yards from the place of use. Here the latite forms the capping of a large landslide block and is tilted about 15° to the west. The lava has the usual black obsidian base, and above, the pinkish or reddish and the still higher grayish platy facies. Through the pinkish zone are layers or bands of fairly massive, non-platy, and only slightly vesicular rock. Rock from the more compact bands has been quarried in two or three places, but all near together. The one highest on the hill is on a particularly pinkish or reddish layer, and from it blocks three feet thick are readily procured with careful blasting. The lava is broken by widely spaced joints, usually as much as five feet apart. Only the upper three feet of the band has been quarried and it is not known to what depth it extends. The rock has a dull, light reddish color; is highly porphyritic with numerous glassy plagioclase phenocrysts in a reddish groundmass. Its structure is minutely vesicular, but not enough to spoil its use as building stone. The rock has also a faint fluidal banding. The layer has been quarried along an area 75 feet long and 20 feet wide. Much more is available, for the thin platy rock above is not thick. Most of the platy facies has been eroded and does not approach serious proportions for a score of feet. Most of the quarried rock, however, is from a point about 100 yards to the west and lower on the hill, and at about the same horizon. The rock has about the same color and texture, although in places it is only faintly pinkish. This part of the flow is cut by widely spaced vertical joints, in places 10 or 12 feet apart, and by several horizontal joints 12 to 18 inches apart. The worked layer is mainly massive or only minutely vesicular. The rock has been described in petrographic detail in another section of the report. The quarry face is about 150 feet long and seven feet high, all in workable rock (Plate XVII, A). The overburden is from nothing to two feet, but will continue to increase as the working face advances. The amount available is large and far beyond local demands. About 15 feet below, the lava changes to the obsidian facies; a short distance above to the grayish latite, which in addition to its platy character has large vesicles and is much too vesicular for use.

Quartz latite has also been utilized to a moderate extent for dwellings and buildings in Oakley. The grade school has a foundation of pale reddish colored quartz latite built nearly eight feet above the ground. The L.D.S. church on the east side of town is constructed wholly from blocks of latite except for cornerstones of white tuff (Plate XVII, B). These blocks have all been trimmed to uniform size and have been aligned in a most attractive manner. Latite is also used as a foundation stone in many of the dwellings in Oakley. Most of the rock has been quarried on Cottonwood Creek, seven or eight miles west of Oakley and outside the area discussed in this report. The same kind of rock is as readily available several miles east of Oakley or in the canyons to the south.

Rock so widely distributed in the area as this should come into more general use as a building stone. As a foundation stone, subject to the action of ground water, or for piering or shoring of any sort, the pinkish latite is far superior to the tuff. It is not as readily dressed, however, and is therefore less suited for the fronts of buildings than the tuff. The color, too, is darker and perhaps less desirable. Yet it can be trimmed into remarkably uniform blocks, as has been done in the stone used in the L.D.S. church at Oakley, and because of its color it really adds to the attractiveness of the buildings. Where the stone buildings are trimmed with rock of other hues, the effect is most striking. Although the cost of this stone may be higher than for other materials, its strength and durability and its resistance to weathering make it highly desirable for permanent structures.

BASALT

Basalt has been rather extensively used as a foundation stone at Declo and to lesser extent in the walls of some of the smaller buildings. It has many of the physical properties possessed by the quartz latite—durability and toughness—but is less attractive because of its dark gray or black color. Used in combination with other stones, it adds more tone and gives most pleasing effects, especially when used as the foundation or chimney stones or when scattered through the walls with other varieties, etc. The basalt flows lack the platiness of the latitic flows, but are generally much more vesicular and have much larger vesicles. For this reason the upper part of the flow is generally unsuited for use in the walls of buildings. Toward the center of the flows the rock becomes more massive and has much smaller vesicles. Most of the basalt has columnar jointing, but, as usual in the Snake River basalts, the columns are thick or the joints widely spaced. By judicious selection, blocks of considerable size may be quarried and these may be trimmed in desired size and shape with no greater difficulty than is experienced in working the quartz latites, although the basalt is somewhat more resistant to hammer blows.

The Snake River basalt is the only basalt that has been used. Its distribution in the northern part of the County has been cited in another place, also its petrographic characters. As it has not been folded or faulted or dissected by gullies, natural quarry sites are in general lacking, but most of the basalt is beneath a very scant overburden. The basalt shows not the least sign of weathering and should last indefinitely. Its use in foundations can be strongly recommended.

With reference to the use of basalt in other parts of the State, Loughlin[1] remarks:

"The Zion Cooperative Mercantile Institution building at Idaho Falls, built in 1884, has a front entirely of basalt in dressed blocks, with both rock face and finely tooled surface; some blocks are six feet long. The stone is of nearly black color and contains several small to large irregularly spaced vesicles. It gives a somber appearance, but shows no weathering effects after 30 years of exposure, and door sills where loaded trucks have been passing back and forth have not suffered any noticeable abrasion. Basalt has also been used, and with good effect, in the base of the Methodist Episcopal church at Lewiston, erected in 1907. Here the blocks are mostly small, with rock face exposed."

QUARTZITE

The pre-Cambrian quartzites afford excellent possibilities for utilization as building stone. The general character and distribution of these rocks are described in another section. They have not been prospected or tested, but they occur in such abundance as to justify extensive development should market and other conditions warrant exploitation. So far as could be learned, only one dwelling has been entirely constructed of quartzite and this one is probably the most pretentious in the County. The house is on the property of the Melcher Mining and Milling Company and has been constructed from quartzite blocks collected on the ground. Since the quartzite has given such excellent results in this dwelling, it should find more general use in the district.

Most of the quartzite is white and nearly vitreous in luster. Bedding planes are commonly spaced about a foot apart. For this reason the quartzite may be quarried in blocks of about the desired shape. Deep quarrying would perhaps not be necessary, for immense reserves are available at the surface, much of it already scattered around in loosened blocks. Such material is available in the north end of the range and not far from the most thickly settled parts of the County. Ledges suitable as quarry sites should not be difficult to find. For use in foundations, piering, etc., the rock has no equal among the natural stones and should last indefinitely.

MARBLE

Marble has been used in the lower part of one of the business buildings in Oakley, white tuff being used for the remainder. Some also has

[1] Loughlin, G. F., U.S. Geol. Survey Mineral Resources for 1913, pt. 2, p. 1379, 1914.

been used locally for tombstones. This rock has been quarried from the crystalline limestone or marble beds in the upper part of the pre-Cambrian Harrison series not far southeast of Oakley. This marble should prove a valuable resource. Blocks are usually small, but some as much as three feet thick and nearly twice as long have been blasted from several of the quarries and no doubt much larger slabs could be obtained with undercutting and proper quarrying methods. Some layers are of exceptional whiteness, much like the fine Georgia marble. Most of it, however, is mottled white and gray, and if blocks were properly cut and polished, it would be a very attractive ornamental stone. All the marble is well crystalline, of medium grain, and in general is mainly free from foreign minerals of metamorphic origin. Little attempt has been made to utilize it as a building stone, but some has been burned to lime for local use. The deposits will be described in greater detail in the section on Limetone and Marble.

OTHER ROCKS

Other rocks might also serve as building stones. For example the Brazer limestone contains thick beds of uniform character, relatively free from chert and other impurities, that might be worked into blocks suitable for building. Some at the north end of the Albion Range might serve for this purpose and has an advantage over many of the other varieties in that it is not far from the main highway and the railroad, and is adjcent to the most thickly populated part of the district. The sandstones in the Wells formation were not examined with the idea of their utilization as building stones, but it is possible that they might suffice. Because of their failure to produce ledges, it is difficult to determine their adaptability to quarrying. Most of the surface blocks are hard, dense, nearly quartzitic, and brown from oxidation. Some of these when broken have gray centers. The suitability of the sandstone can be determined only by prospecting.

Granite does not give much promise as a building stone. That now exposed at the surface in the City of Rocks area is much shattered and weak and if quarried and used as building stone would soon disintegrate. This shattering or granular disintegration probably extends many feet below the surface or into the exposed ledges. But even though the outer disintegrated shell were removed the rock would not possess the desired qualities of a building stone or ornamental stone because of the white opaqueness of its feldspars. This alteration of the feldspars is not due to surface weathering, but is probably the result of deep-seated action soon after the rock had solidified. For this reason the rock would show no improvement with depth or assume the qualities especially demanded in monumental stones.

LIMESTONE AND MARBLE

Limestone and marble are among the most abundant and most widely distributed mineral resources of the region. Mention has already been made of the use of some of the marble as building and ornamental stone and of the possible use of the Brazer limestone as building stone. Aside from these uses, the rocks are also adapted for burning to lime, as a cement material, as agricultural lime, chemical lime, and perhaps other purposes.

CARBONIFEROUS LIMESTONES

The Brazer formation and the lower part of the Wells formation contain many beds that appear to consist of relatively pure limestone, much of which may be classed as "high-calcium" lime; that is, which carries more than 93 per cent calcium carbonate. The distribution of these rocks has already been discussed and may be seen by reference to the geologic map. On the whole, the Brazer beds appear to be more nearly pure than the others, though this formation includes beds of sandstone and shale in addition to limestone. Some of the limestones were analyzed in the Idaho Bureau of Mines and Geology laboratory[1] and the results obtained are given in the following table:

	1	2	3
Insoluble (silica)	1.35	8.95	1.63
Calcium carbonate	94.8	91.3	95.7
Magnesium carbonate	3.66	2.37	1.30
Iron oxide	.21	.29	.14
Aluminum oxide	.17	.3	.18

1. Sample from massive light gray limestone (Brazer), southern end of the Black Pine Range in Sec. 9, T. 16 S., R. 29 E.

2. Sample of the Brazer limestone in Sec. 7, T. 12 S., R. 30 E., in Power County.

3. Sample from the Lower Wells formation on the west slope of the Albion Range, east of Oakley.

The sample from the Black Pine Range (1) is from the ledge of white or light gray massive limestone so conspicuous and widely distributed in that part of the Range. It is much fractured and seamed with calcite.

---

[1] Analyses by Mr. R. V. Lundquist, chemist, Idaho Bureau of Mines and Geology.

As this limestone contains more than 93 per cent of calcium carbonate, the product obtained by proper burning would be classed as high-calcium lime. The same member or horizon is also widely distributed in the northern end of the Sublett Range and probably in the north end of the Albion Range. It is reported that analyses made on limestone from the north end of the Albion Range yield consistently less than one per cent magnesia, but usually more than 10 per cent silica or insoluble. Sample (2) from the Sublett Range is from a grayish, dense, fine-grained limestone containing numerous cup corals and crinoid stems. The silica or insoluble material is comparatively high and the product obtained on burning would fall below the grade of high-calcium lime. The sample, however, is not wholly representative of the bed. Sample (3) from the west slope of the Albion Range east of Oakley is from a dark gray limestone in the lower part of the Wells formation. This limestone would also yield high-calcium lime. Its low content in magnesium carbonate and silica is worthy of note.

None of the Carboniferous limestones have been utilized. It is certain, however, that limestone of suitable quality exists in sufficient variety and quantity to form a potential resource of great future value. Much of the limestone is suitable as a cement material. It may also be burned to lime and used for many purposes, such as building lime, finishing lime, agricultural lime, and perhaps chemical lime. The utilization of the limestone will probably be slow, in view of the great amount of this material more readily available or nearer railroads in southeastern Idaho. There probably will be no demand for the lime as a cement material so long as a cement plant continues in operation near Pocatello, but the limestone for this purpose must be regarded purely as a potential resource.

PRE-CAMBRIAN MARBLE

The marble beds in the middle division of the Harrison series also possess many desirable qualities. Much of the rock appears to consist of relatively pure calcium carbonate and is of high-calcium rank, although the magnesium content is somewhat higher than in most of the Brazer limestone. The general distribution of these rocks along the western slope of the Albion Range has been discussed in another place. The marble has been quarried in a number of places and the rock burned locally to lime and to some extent used at the sugar factory at Twin Falls. Samples from several of the quarries were analyzed and the results are recorded in the following table (R. V. Lundquist, analyst):

	1	*2*	*3*	*4*
Insoluble (silica)	5.25	1.06	0.50	0.77
Calcium carbonate	87.7	94.0	97.4	97.4
Magnesium carbonate	4.85	4.57	1.38	1.05
Iron oxide	.79	.21	.14	.14
Aluminum oxide	.66	.19	.09	.09

1. Quarry at north end of the Albion Range in Sec. 31, T. 10 S., R. 25 E.
2. Quarry near highway between Albion and Declo in Sec. 24, T. 11 S., R. 24 E.
3. Quarry near the mouth of Smith Creek, Sec. 24, T. 12 S., R. 23 E., grayish mottled bands.
4. Same quarry as (3), but of the white marble bands.

The marble has been exposed in several small quarries at the very end of the Albion Range about three miles east of Declo, less than one-fourth of a mile from the main highway, in Sec. 31, T. 10 S., R. 25 E. An analysis of the marble from one of the small quarries is given in Column 1. This agrees closely, except for the high percentage of insoluble, with analyses furnished by the owner of the property. It is reported that the marble contains up to 11 per cent magnesia, but the average is between 3 and 5 per cent. It is also reported that 500 tons were used by the sugar factory at Twin Falls. The marble is mainly thin-bedded and unsuited for building stone. It is rather coarsely crystalline, grayish, mottled, and has some disseminated pyrite and lime silicates. About 15 feet of beds are exposed in the quarry, but because of overburden the true thickness of the member could not be determined. The beds strike about N. 55° E. and dip about 20° NW. The same series of beds are again exposed several hundred yards to the northeast, but little rock has been quarried at that point. The marble is overlain by schist. No doubt the marble beds appear in many other places in this part of the range.

Marble has also been quarried along the highway between Albion and Declo near the upper part of the grade where it makes rapid descent from the summit of the range to the plain below. The beds appear much thicker here than farther north and measure more than 60 feet across in the working face. The marble is massive and bedding is very difficult to distinguish. Most of it is moderately coarsely crystalline, free of pyrite or other minerals, grayish, mottled and banded. It should prove suitable for monumental and building stones, provided blocks of sufficient size can be obtained. At or near the surface the marble is much shattered, although blocks two or three feet wide and five or six feet long, free from cracks, may be seen in the quarry side. With depth the cracks should become less numerous and more widely spaced and eventually disappear altogether, affording rock of suitable size and quality. Analysis of this rock is given in column 2. It may be classed as high-calcium lime with nearly 5 per cent magnesium carbonate. The deposit has been opened by a cut, glory hole, and a short tunnel and shaft about 60 feet below the outcrop. This marble has been burned to lime, and remains of an old kiln may still be seen. Apparently the plant has not been operated for many years. The beds strike about N. 85° E. and apparently dip about 55° SE. into the hill. They may be traced several hundred yards to the southwest by the line of float, but beyond that point the marble is effectively concealed by detritus from above. Its actual width could not be determined. Reserves at this place are large and probably far above local demand.

Marble has also been quarried on Smith Creek about three and one-half miles southeast of Hazel, probably in Sec. 24, T. 12 S., R. 22 E. very near the base of the Albion Range. The structural relations here are very complex, as there appear to be several beds of marble separated by quartzites; all much disturbed by faulting. Two glory holes have been made on the main bed of marble, which lies to the east of a fault which has brought these beds in contact with a different part of the series of marbles on the other side. This marble is moderately coarsely crystalline, grayish, mottled, but has some layers of pure white. Much of the marble exposed in the two glory holes is broken into small blocks by jointing and weathering. Some blocks on the waste pile are as much as four feet long, three feet wide, and two feet thick. Blocks of size suited for monumental stone may be obtained, and the proportion of large blocks is likely to increase with depth below the surface zone of joints and fractures. Chemical analyses were made of both the mottled and the white marble, with the results recorded in columns 3 and 4. The high-calcium content of this marble is at once apparent. The bed is about 100 feet thick and its trend as measured in the two glory holes is about N. 75° E. and dip 40° NW. The bed may be traced across two ridges and then intermittently for a number of miles. Immediately to the west of the glory holes the beds are cut off by a fault (outlined by the gully) and lie against beds of impure gneissic marble belonging higher in the series. These strike about N. 20° E. and dip 30° SE. The marble is much weathered and stained from the alteration of some of its silicate minerals. In places the strike changes to the west and locally is as much as N. 60° W. About 100 feet of these beds, the upper part composed of relatively pure marble, show in the hill beneath the capping of quartzite. The bed may be traced to the north for more than two miles. Much of the marble is not of sufficient purity to interest development. Several zones of marble, apparently the entire upper series, appear on the east side of the fault at the top of the ridge north of the glory holes. One of these is continuous with the marble bed exploited below and has a thickness of about 105 feet. This lies beneath about 105 feet of the impure gneissic marble. Then follows about 75 feet of massive grayish marble, 30 feet of gneissic marble, 25 feet of white and mottled marble, and finally quartzite.

On the south side of Robinson Creek some of the marble forms a ledge about 85 feet high, overlain and underlain by quartzite. Another bed of marble about 30 feet thick outcrops higher on the ridge. The same beds exposed on Smith Creek appear also on Willow Creek and other streams to the south, especially on Boulder Creek and Land Creek. The marble beds are several in number and in the aggregate total several hundred feet. Some of the marble on Land Creek has been burned for lime.

Marble also outcrops extensively southeast of Basin or Oakley, beginning near the foot of the Albion Range near Spring Creek and Slide Creek. It is the grayish and white kind so widely distributed farther north and may be quarried in blocks of considerable size. Some has been utilized both for stone and for lime. This marble is readily accessible and easily quarried. Outcrops also appear south of Mill Creek, but the deposits do not appear south of North Carson Creek. The last quarry is in Sec. 28, T. 14 S., R. 23 E. The marble is cut by the granite, but no contact lime silicates were noted. It is mainly white, granular, and distinctly bedded. The member is probably more than 50 feet thick and has been prospected in several places by small cuts. Some of the marble has been burned to lime, for an old lime kiln still stands on the premises.

It seems certain that marble of suitable quality exists along the west margin and north end of the Albion Range in sufficient variety and quantity to form a potential resource of considerable value.

QUARTZITE

The quartzite in the Harrison series is ordinarily rather pure and vitreous, and should prove a valuable resource in addition to its use as a building stone. Some of it has all the qualities possessed by vein quartz and may be crushed or ground and used in a variety of ways.[1] Pure quartzite may be used in the manufacture of pottery, in metallurgical and chemical processes, as an adulterant in paints, as a filler for wood, in the manufacture of silica glass, as a road filler, as an

---

[1] Loughlin, G. F., U.S. Geol. Survey Mineral Resources for 1913, pt. 2, chapter on silica, 1914.

abrasive in scouring soaps, in making sandpaper, in the manufacture of polishing powders, for "frosting" glass, and for other purposes. As an abrasive, crushed quartz is preferable to sand or crushed sandstone because it has sharp edges. For the same reason it is preferred for paints and as wood filler, because grains with sharp edges adhere more closely than those with rounded edges.

The quartzite in the northern end of the Albion Range is favorably situated with respect to transportation, both by highway and by railroad. Location of quarries involves only the selection of those places with a minimum amount of overburden and easy access to the highways. The quantity of rock in the Albion Range needs no estimation, for the supply is inexhaustible. Utilization of this resource can come with the establishment of abrasive, ceramic, metallurgical, and chemical industries in the surrounding region. It probably would be impossible to ship the quartzite to eastern markets in competition with deposits nearer the point of demand, because of the long, expensive haul across the country.

VOLCANIC ASH

Volcanic ash or pumicite is a widely distributed resource of considerable potential value. Its general characters and distribution have been discussed elsewhere, much of the so-called tuff in the Tertiary series being in reality thick beds of white to light gray, fine to coarse, pumiceous ash of rather exceptional purity. High grade pumicite has a variety of uses, especially in abrasives. Usually it is ground so as to give a uniform product, when it may be used as the principal ingredient in many cleansing compounds, in abrasive hand soaps, in metal polishes, tooth pastes, and as a filler for paints. It has also found use as a lightweight concrete aggregate, taking the place of a considerable part of the cement.

None of the pumicite has so far been utilized. Its exploitation is controlled by the same factors that govern the marketing of the quartzite—namely, the long distance from present markets where the supply is from much nearer sources. Much of the ash is concealed beneath the Tertiary lavas, but exposures are sufficiently numerous in the fault scarps and deeper gullies to provide an immense reserve of this material. Deposits are farther removed from the railroads than the quartzites, but this factor is more than balanced by the cheaper costs in quarrying and grinding. Adequate exposures are on the east and west flanks of the Malta Range and in the canyons southwest and south of Oakley. Deposits in the Sublett and Black Pine ranges are too far removed from railroads to compete successfully with those in other parts of the district.

MICA

Considerable prospecting has been done for mica near the southern end of South Mountain, but so far none of commercial size has been found. Such muscovite as has been found forms small plates or books, rarely more than four inches across and seldom more than two, scattered widely through the pegmatite bodies. These lie in any position in the pegmatites and show no tendency to form nests or seams. With such an abundance of pegmatites in some parts of the region and with so many holding muscovite plates, there is always the possibility that some with marketable sizes of mica may be found. But the failure of large books to show in any of the outcrops or in any of the surface debris does not lend encouragement.

FELDSPAR

Feldspar is an abundant constituent in the pegmatites admixed with quartz, muscovite, biotite, and several minor accessories, as discussed in the section under igneous rocks. Feldspar is employed mainly in the ceramic industry and to a less extent as an abrasive in scouring soaps and window washes. It is also used in the manufacture of glass, roofing material, and concrete. Most whiteware bodies consist of 10 to 35 per cent of feldspar and glazes usually contain from 30 to 50 per cent of this material.

As the pegmatite bodies are usually small and irregular, the mines must also be small and irregularly developed and operated. Few deposits farther than five miles from a railroad can be operated successfully. Unless exceptionally well located or of unusual purity, deposits less than 25 feet wide are of little commercial interest. For economical operation, marketable feldspar should constitute at least 50 per cent of the rock mass; should occur in fairly large clean masses, so that sorting is easy; should be free from iron stains, and other finely disseminated impurities; and quartz should constitute not over 25 per cent of the shipping product.

Most of the pegmatites are in the southern end of South Mountain, nearly 18 miles from the end of the railroad at Oakley. Some of the bodies are large, exceeding 25 feet in thickness, and may be traced for several hundred feet on the surface. In some the quartz and feldspar are intricately intergrown as graphic granite, composed normally of about 75 per cent feldspar and 25 per cent free quartz, but most of them have essentially the textures of very coarse grained granites. In most of them the feldspar constitutes nearly half of the rock. Should there ever be a demand for feldspar from the general region, no doubt suitable deposits may be found for exploitation.

CLAY

No special study was made of the clay occurrences in Cassia County. Considerable clay has been mined and burned to bricks in a plant at the east edge of Burley. A good grade of brick is produced and is used extensively in buildings and houses in Burley and in the surrounding region. This clay is of the transported kind and is mined from the silt or floodplain deposits not far from the banks of the Snake River. Reserves are probably sufficient to supply the local demand indefinitely. Clay has also been burned in other parts of the district, but none other than the plant at Burley is in operation at the present time.

High grade clays or kaolins were not seen anywhere in the district, but the lower grade suitable for structural uses are widespread. In addition to the clay or silt deposits along the Snake River, there are extensive deposits of aeolian clays or loess which are equally as well adapted for burning to ordinary bricks or tile. Such deposits have been mentioned elsewhere as occurring north of the town of Albion and in the north end of the Sublett Range and the bordering Snake River Plains. In the latter locality the aeolian deposits are exceptionally thick and easily mined, but will never find utilization in Cassia County, because deposits of the same kind are more easily available along the railroad at American Falls in Power County.

CYANITE

Cyanite has been found in the north end of the Albion Range, in Sec. 9, T. 11 S., R. 25 E., in boulders or blocks derived from the thick garnetiferous schist member in the upper part of the middle division of the Harrison series. Several small boulders have been found on the surface, in part resting on the schist and in part in the talus slide below the outcrop. These boulders or blocks contain small lenticular masses of bladed crystals or crystal aggregates measuring from an inch or two thick to nearly a foot. They constitute relatively pure cyanite aggregates whose individual crystals of pale bluish color form blades as long as six to eight inches and an inch or more wide. The nests are firmly bedded in the highly garnetiferous schist and some of the blades penetrate into the schist. Quartz is commonly associated with the cyanite, and in some there are large flakes of brownish phlogopite. Microscopic studies of the enclosing schist show it to be composed essentially of sillimanite, with some andalusite, quartz, graphite, garnet, and several minor accessories. Relict structures of the schist may be seen in the cyanite nests without the aid of the microscope and these relations suggest that the cyanite has been developed by the recrystallization of the materials in the schist or by replacement of these materials. The cyanite masses are not readily detached from the firmly held schist.

Cyanite contains the same chemical elements as andalusite and in the same proportions, and should serve the same purpose as andalusite. The latter is being used in the manufacture of sparkplug porcelains, and if it could be produced cheaply it might be used as a source of alumina in many other ceramic processes. Not only is there a demand from the potters for aluminum material, but it has been learned that aluminous refractories are much more satisfactory than firebrick for certain purposes.

No search has been made to find the part of the schist in which the cyanite is developed. Cyanite is a frequent or common constituent in metamorphic rocks, although it rarely forms large masses or occurs in such quantity as to encourage commercial development. It is very unlikely that it will be found in any greater abundance than in occasional scattered nests and then probably very locally. This schist member may be traced for many miles along the west margin of the Albion Range, but so far as known, cyanite has not been found in any part of it except in the one particular locality. Its localization is perhaps the result of contact action in the neighborhood of granite, although no granite appears at the surface in this part of the range. It should require very little effort to find the cyanite in place, as most of the schist (here about 200 feet thick) projects above the surface or is beneath a comparatively thin overburden. Its interest will probably prove to be more scientific than commercial.

ROAD METAL

Supplies of rock suitable for road metal are very abundant in Cassia County, but comparatively little use of them has been made. Most of the surfacing of roads has been with gravels derived largely from the same formations which are considered best adapted for use after crushing. Obviously, it is much less expensive to utilize rock already broken than to quarry and submit the material to crushing. While crushed rock possesses some qualities on the road lacking in rounded gravel, these same qualities may be obtained by submitting the gravels themselves to crushing. Should crushed rock be desired, practically inexhaustible supplies are readily available. Such rocks adapted for use as road metal include the Snake River basalt, the Tertiary latites, pre-Cambrian quartzites, and, in some of the more remote parts of the area, the Carboniferous limestones and cherts.

Where rock is intended for water-bound macadam construction, the most essential qualities to be determined are hardness (resistance to abrasion), toughness (resistance to impact of traffic), and binding power or cementing value (ability of the rock powder when in contact with water to bind or cement the larger rock fragments and prevent their displacement under the shearing action of traffic). In general, the igneous rocks possess qualities superior to those of sedimentary rocks. In south Idaho the cementing value of the crushed rock or gravel has been one of the critical factors in the construction of permanent macadam roads. No rock has yet been found that would stand under the arid climatic conditions, but binding material is soon loosened during the dry season and carried away by the winds. This difficulty in road construction has recently been overcome by the use of a bituminous or oil binder.

Crushed Snake River basalt is probably the best material for permanent roads. Usually it is superior in hardness and toughness, and its cementing value, especially in the slightly weathered rock, is high. This rock is available across the northern part of the district in the line of the heaviest traffic, but none of it has been utilized. The rock is exceedingly fresh and none of it has been eroded to form gravel deposits. It would therefore be necessary to quarry and crush the rock from the ledge before it could be used on the road and it would probably be necessary to add an oil binder to obtain the best results under the arid climatic conditions.

The Tertiary quartz latites also possess desirable qualities as road metal. As much of the latite has a platy character, it would probably reduce to smaller fragments much more readily than the basalt, but also would probably break down more easily under heavy traffic. No doubt its wearing or abrasive qualities would also be somewhat less. None of it has been quarried and crushed directly from the ledge, but some obtained from talus slopes has been used on certain of the secondary roads. It appears to be satisfactory under the lighter traffic and its value could probably be improved by removing some of the larger fragments or blocks from the roadways. Gravels derived from the latites have been used on the South Branch of the Old Oregon Trail near Idahome. The gravel cements better than most of the other materials used in the district, but the road becomes dusty and ravels in the dry, hot part of the summer. The rock is available over much of the area, either in the ledge or in the gravels at the base of the lava ridges. Crushed talus or gravel would serve as well as the crushed massive rock and would prove much less expensive. Such material would also possess better cementing qualities because of the weathered materials and clays or silts admixed with it.

Quartzite, which is composed essentially of the hard mineral quartz and in this region is a recrystallized sandstone of much the same character as vein quartz, is superior to all other sedimentary rocks and equal to or superior to most igneous rocks in resistance to wear, hardness, and toughness, but is inferior to all in cementing value. But where bituminous material forms the binder, the cementing value of the rock used as surfacing material is of less importance. Such quartzite can well suffice in or near the Albion Range, the Raft River Range, and South Mountain, but is not available elsewhere except in some of the transported gravels or extensive alluvial slopes at the base of these ranges.

Rocks available in the eastern part of the County are mainly the Carboniferous limestones, sandstones, and cherts, locally latite and basalt. Sandstone is not satisfactory as a road material, as it is readily reduced to sand under light traffic. Chert is locally abundant along Sublett Creek and it might be utilized, especially as some of the outcrops of the Rex chert member are well located and much fractured, so that quarrying would be easy. Its wearing qualities would perhaps be lower than for sandstone and limestone, but its hardness and toughness should exceed both. It should also develop good binding qualities on the road. Limestone has in general an excellent cementing value, but its hardness is low and the wearing qualities only fair. It is suitable under light traffic. It is probably the best material available in the southeast part of the County and may in places be utilized from the gravel deposits which extend from the base of the Black Pine Range.

Granite is not well fitted to serve as a road dressing. Much of it is so shattered or disintegrated that it would readily reduce to a sand under traffic. However, with a suitable base beneath, it could serve as a top dressing and would yield a smooth sandy road. Its utilization could be only local on roads mainly in the City of Rocks area.

SCENIC RESOURCES

The scenic resources of Cassia County could be much more fully utilized. In the Cassia City of Rocks the district has one of the most unique features of its kind in the country and perhaps in the world.

Although picturesque erosion features, such as rather striking displays of large boulders, domed monoliths, castellated forms, and pedestal rocks of varied size and shape, are fairly common in the granitic rocks of arid and semi-arid regions, in no place, so far as the writer knows, except possibly in the Buffalo Mountains in Australia, are they so strikingly developed and brought together within such a relatively compact area as in Cassia County. The rock city is within forty miles of the Old Oregon Trail and could be made an interesting side trip for the thousands of tourists who cross the northern end of the district and know nothing of its presence. The old immigrant road to California passed about a mile to the south of the main part of the rock city, and needless to say, the City of Rocks was better known to the pioneer travelers who first settled southern Idaho than to residents of Idaho today. It is a scenic resource of great worth, as yet practically undeveloped. Its features are sufficiently unique and interesting so that they should be included in a national playground or National Monument. No comment is necessary as to the value of such a resource to Cassia County were it widely advertised. The thousands of tourists who would enjoy seeing these spectacular features and who could be made to linger in the region is answer enough.

INDEX

PUBLICATIONS OF THE IDAHO BUREAU OF MINES AND GEOLOGY

MOSCOW, IDAHO

BULLETIN No. 1—The Copper Deposits of the Seven Devils and Adjacent Districts.*

BULLETIN No. 2—A Preliminary Report on the Clays of Idaho.*

BULLETIN No. 3—A Reconnaissance in South Central Idaho, price 15 cents.

BULLETIN No. 4—Petroleum Possibilities of Certain Anticlines in Southeastern Idaho.*

BULLETIN No. 5—Geology and Ore Deposits of Alturas Quadrangle.*

BULLETIN No. 6—Geology and Water Resources of Goose Creek Basin.*

BULLETIN No. 7—Geology and Gold Resources of North Central Idaho, price 50 cents.

BULLETIN No. 8—Geology and Oil Possibilities of Bonneville, Bingham and Caribou Counties, price 50 cents.

BULLETIN No. 9—Geology and Ore Deposits of Boise Basin, price 50 cents.

BULLETIN No. 10—Geology and Ore Deposits of Boundary County, price 50 cents.

BULLETIN No. 11—Geology and Metalliferous Resources of the Region About Silver City, Idaho, price 50 cents.

BULLETIN No. 12—Geology and Ore Deposits of the Clark Fork District, price 50 cents.

BULLETIN No. 13—Craters of the Moon National Monument, Idaho.*

BULLETIN No. 14—Geology and Mineral Resources of Eastern Cassia County, Idaho, price 50 cents.

PAMPHLET No. 1—Interfacial Tension Measurements and Some Applications to Flotation.

PAMPHLET No. 2—Size of Mineral Particle in Relation to Flotation Concentration.*

PAMPHLET No. 3—Testing Ores for Flotation.*

PAMPHLET No. 4—Differential Flotation.*

PAMPHLET No. 5—A Preliminary Reconnaissance of the Gas and Oil Possibilities of South Central and Southwestern Idaho.*

PAMPHLET No. 6—A Preliminary Study of Certain Reported Platinum Occurrences near Coeur d'Alene, Idaho.

PAMPHLET No. 7—Notes on the Geology of Eastern Bear Lake County, with Reference to Oil Possibilities.*

PAMPHLET No. 8—Ground Water Supply at Moscow, Idaho.*

PAMPHLET No. 9—Ground Water in Pahsimeroi Valley, Idaho.

PAMPHLET No. 10—The Horseshoe Basin Area of the Teton Coal Field in Southeastern Idaho.*

PAMPHLET No. 11—Geology and Water Resources of the Bruneau River Basin.*

PAMPHLET No. 12—Possibilities of Petroleum in Power and Oneida Counties.

PAMPHLET No. 13—A Geologic Reconnaissance of the Mineral and Cuddy Mountain Mining Districts.

PAMPHLET No. 14—Mica Deposits of Latah County, Idaho.

PAMPHLET No. 15—Ground Water for Irrigation on Camas Prairie.

PAMPHLET No. 16—Ground Water for Municipal Supply at Idaho Falls, Idaho.

PAMPHLET No. 17—Ground Water for Municipal Supply at St. Maries, Idaho.*

PAMPHLET No. 18—Some Miocene and Pleistocene Drainage Changes in Northern Idaho.*

---

* Publications out of print.

*Publications out of print.

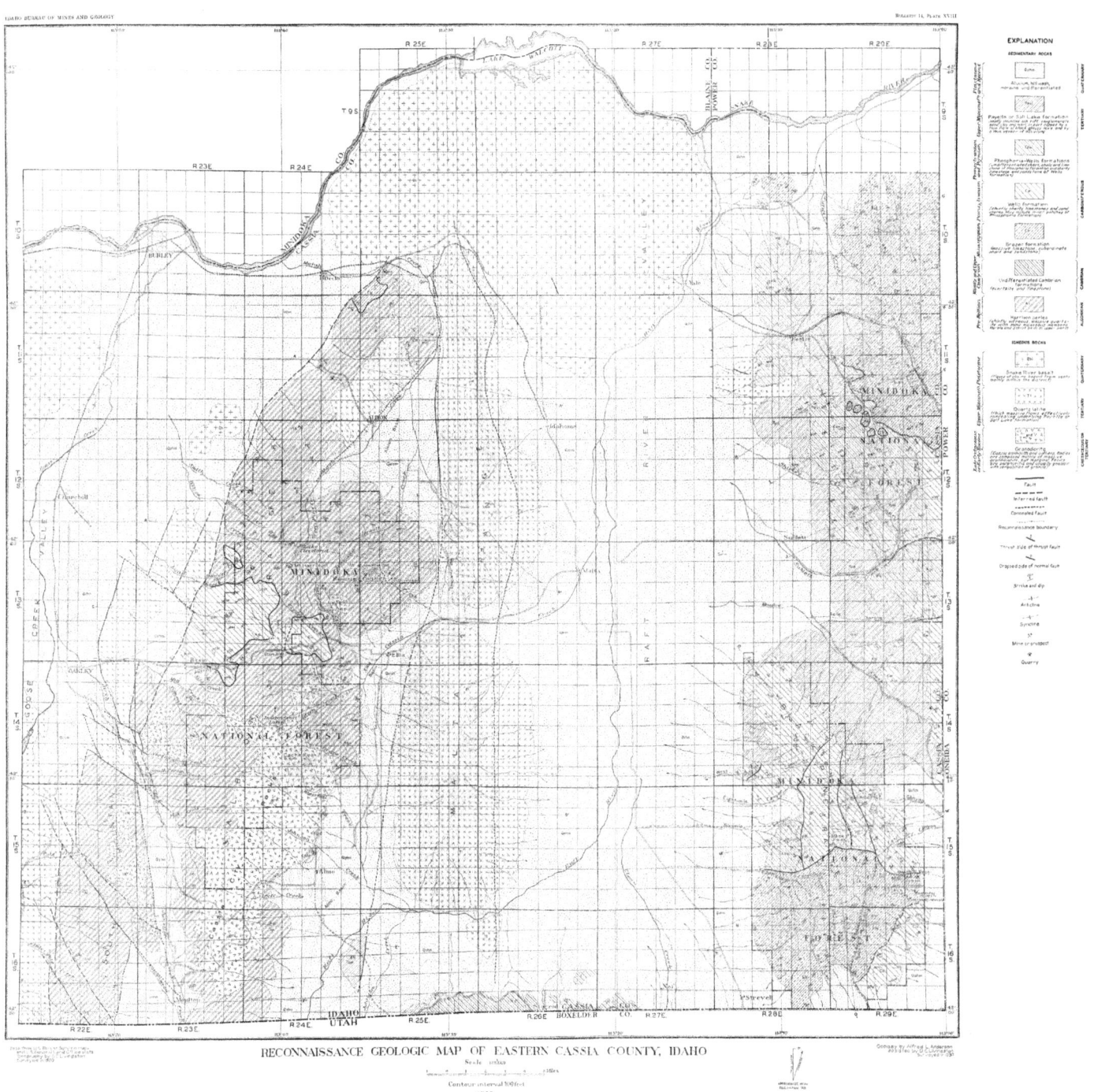

RECONNAISSANCE GEOLOGIC MAP OF EASTERN CASSIA COUNTY, IDAHO

Scale 1:125000

Contour interval 100 feet

1930

EXPLANATION

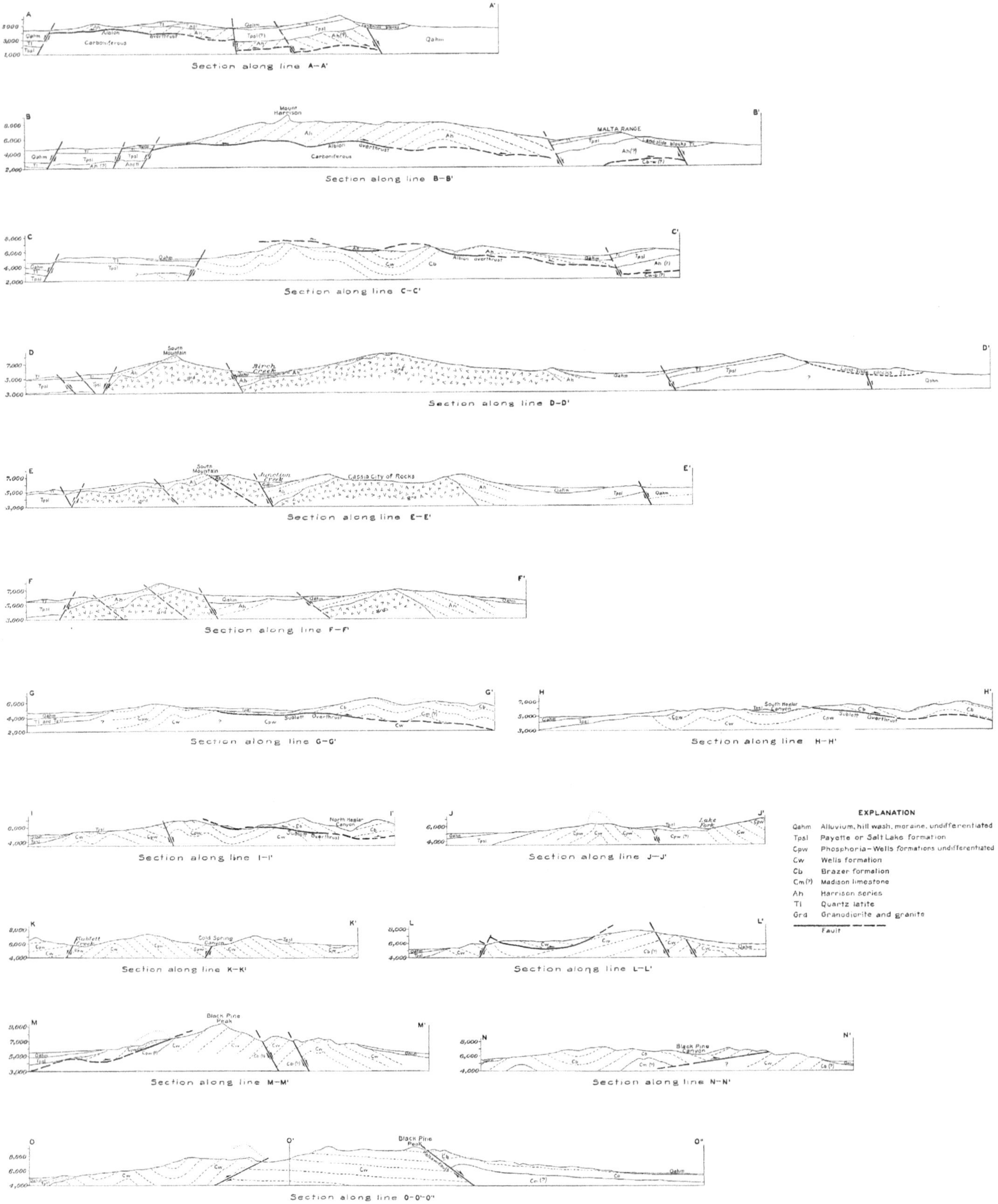
Section along line A–A'
Albion overthrust
Carboniferous
Section along line B–B'
Mount Harrison
MALTA RANGE
Landslide blocks
Section along line C–C'
Albion overthrust
Section along line D–D'
South Mountain
Birch Creek
Landslide blocks
Section along line E–E'
South Mountain
Junction Creek
Cassia City of Rocks
Section along line F–F'
Section along line G–G'
Sublett overthrust
Section along line H–H'
South Heglar Canyon
Sublett overthrust
Section along line I–I'
North Heglar Canyon
Sublett overthrust
Section along line J–J'
Lake Fork
Section along line K–K'
Sublett Creek
Cold Spring Canyon
Section along line L–L'
Section along line M–M'
Black Pine Peak
Section along line N–N'
Black Pine Canyon
Section along line O–O'–O''
Black Pine Peak
EXPLANATION
Qahm Alluvium, hill wash, moraine, undifferentiated
Tpsl Payette or Salt Lake formation
Cpw Phosphoria–Wells formations undifferentiated
Cw Wells formation
Cb Brazer formation
Cm(?) Madison limestone
Ah Harrison series
Tl Quartz latite
Grd Granodiorite and granite
Fault

GEOLOGIC STRUCTURE SECTIONS, CASSIA COUNTY, IDAHO

A. CHARACTERISTIC VALLEY OF SOUTH MOUNTAIN AND THE ALBION RANGE
Valleys of the two ranges are characteristically wide and shallow, typically U-shaped, except on the westward slopes. Courses are straight. These valleys are presumably the result of an early epoch of glaciation.

B. EROSION SURFACE ON THE SUMMIT OF THE ALBION RANGE
Old erosion surface preserved on the summit of the Albion Range east of Mount Harrison. This surface lies at an elevation above 8,000 feet A.T. and is cut across the folded pre-Cambrian Harrison series. This surface was formed prior to mid-Miocene time and is now probably mainly an exhumed surface. It is probably equivalent to the Snowdrift peneplain defined by Mansfield in southeastern Idaho.

A. MOUNT HARRISON

Mount Harrison has a broad summit surface lying nearly 1,000 feet above the Snowdrift (?) peneplain which is shown in the immediate foreground. Its appearance suggests a still more ancient erosion surface, but more than likely it is a part of the Snowdrift surface elevated through recent normal faulting. Pleistocene glaciers have carved cirques on its eastern side. Note the broad, gently undulating surface of the Snowdrift (?) peneplain.

B. CACHE PEAK FROM THE NORTH

Cache Peak, the highest point in Idaho south of the Snake River Plains. Note the two wide rock terraces on both sides of the main peak. These are the remnants of the Snowdrift (?) peneplain preserved in the southern section of the Albion Range.

A. GLACIAL VALLEYS ACROSS THE ALBION RANGE

These U-shaped valleys extend entirely across the southern end of the Albion Range from the wide basin between South Mountain and the Albion Range. Ice presumably filled the basin in front of South Mountain and overflowed across the Albion Range into the Almo Basin in the foreground. The Cassia City of Rocks is hidden behind the row of peaks. Cache Peak on the extreme right. Vegetation expresses the general aridity of the region.

B. LAKE CLEVELAND

Lake Cleveland is a small glacial lake lying in a cirque of late Wisconsin age on the northeast side of Mount Harrison in the Albion Range. Note the general smoothness of the old erosion surface on the right, whose general level is but little above that of Lake Cleveland.

A. BLACK PINE RANGE

Shows the three-segment character of the Black Pine Range, with the southern end (right) plunging to the level of the desert plain. The central part rises to 9,700 feet A.T. The deep notch between the middle and right segment outlines the course of the Kelsaw fault, that on the left side an overthrust fault. Picture also shows the southern end of the Raft River Valley, with its monotonous flatness of surface and characteristic vegetation. Distance across to the Black Pine Range is eleven miles.

B. DETAILS IN THE BLACK PINE RANGE

Shows the east side of the middle segment, with its deep, steep valleys and sharp ridges, quite unlike the wide, shallow valleys and wide intervening ridge tops of the Albion Range. In the foreground are the exhumed triangular facets of a normal fault. Pole Canyon follows the fault from the left, but in the center of the view turns off at right angles. Saddle outlining the fault continues ahead. The rocks belong to the Wells formation. Note absence of outcrops on these exceedingly steep slopes. The Wells invariably gives smooth slopes.

A. SUBLETT RANGE

This view in the Lake Fork drainage area shows the advanced stage of dissection of the Sublett Range. In the picture are recorded the cycles responsible for its dissection: the Gannett erosion surface (g) on the highest ridges; the Elk Valley (e) below; the Dry Fork (d) at a still lower level; and the steep valleys of the Blackfoot cycle (b). The Gannett surface is probably late Pliocene in age, the others Pleistocene. These surfaces are carved in rocks belonging to the Wells and Phosphoria formations. Note the absence of outcrops.

B. VIEW OF THE NORTH END OF THE SUBLETT RANGE

The wide valley surface (d) at the level from which the picture was taken records the Dry Fork cycle. Picture was so taken that the steep, narrow canyons of the Blackfoot cycle fail to show. At higher levels may be observed the rock terraces corresponding to the Elk Valley (e) and Gannett (g) surfaces.

A. CASSIA CITY OF ROCKS

General view of a part of the Cassia City of Rocks showing some of the domed monoliths and other forms
near the lower margin of the curious granite outcrops.

B. DOMED MONOLITH

One of the numerous monoliths in the rock city. Such bodies rise steeply above the sandy floor of the
basin and are usually free from coarse fragmental debris at their bases.

A. SOUTH ENTRANCE TO THE ROCK CITY
These bodies keep watch in front of the silent rock city.

B. THE THREE SENTINELS
This shows the marked resemblance of some of the forms to the rock walls in the famous Zion Canyon in Utah.

A. THE ORIENTAL TEMPLE

Some of the forms resemble Oriental temples and mosques. One can easily imagine this area as belonging to a fable city in The Arabian Nights. Note the horizontal jointing in the granite.

B. OLD WOMAN ROCK

Many of the curious forms resemble animals, some human beings. Most interesting of these is the old woman in the picture above. Note the Grecian frieze effect.

A. GIANT TOADSTOOL

Shows the peculiar result of weathering of the granite—removal of disintegrated rock from beneath an upper indurated crust.

B. PEDESTAL ROCK

Perched high on the tapering stem is a large block, hollow on the inside. Weathering has continued inside the block beneath the surface-hardened shell. The rock grains loosened by weathering have dropped to the ground, leaving only the outer shell.

A. EXPOSURE OF BRAZER FORMATION IN THE BLACK PINE RANGE

The white rock is the massive white or light gray limestone in the Brazer formation overlying black carbonaceous shales and sandstone. Most of the ore deposits in the Black Pine Range are in this limestone. Silver Hills mine in the lower foreground.

B. CHARACTERISTIC OUTCROP OF THE BRAZER FORMATION

The Brazer formation tends to form pronounced ledges in conspicuous contrast to the smooth slopes produced by the sandstones of the Wells formation.

A. DETAILS IN THE WEATHERING OF THE GRANITE

Shows the slight exfoliation of the granitic rock, also the darker colored iron-stained surface crust on the top of the rock. Note how the rock unprotected by the hardened crust has dropped away, leaving niches and an overhanging roof. Hydration is believed to be responsible for the weathering of the granite, promoting deep granular disintegration, and rapid surface evaporation and casehardening which preserves the top crust.

B. ROCK HOLLOWS

Deep hollows in the sides of the large rock masses are interesting features. These are produced through removal of the granular disintegrated rock from beneath the outer indurated shell above. Note the uneven or fretted surface of the granite block, as though etched by some powerful solvent. Casehardening again explains this peculiar surface, a chemical and physical induration.

A.   HOLLOW BOULDER

As a result of the granular disintegration and surface induration large hollow boulders are of frequent occurrence. Some have only small openings beneath and open upwards into cavities large enough to hold a man. Loosened grains have dropped from the granite under gravity and in time have caused large openings to form.

B.   BATH TUB ROCK

Many of the large monoliths with wide flat roofs are characterized by basins or "bath tubs." Some of these are as much as four feet deep. These depressions have been worn through the casehardened shell by the rains or water running from the irregular rock surface. The deep granular disintegration has favored the development after a start has once been made.

A.   COLUMNAR JOINTING IN QUARTZ LATITE FLOW

Columnar jointing is beautifully developed in parts of the latitic lava flows.   Elsewhere a thin platy parting is more characteristic.   This shows in a gully near the west base of the Malta Range.   Note the tilted position of the flow.

B.   MALTA RANGE FAULT SCARP

Shows the steep eastward-facing scarp near the north end of the Malta Range.   Shows also the manner in which the range disappears at the extreme right by plunging beneath the level of the Snake River Plains. At least two flows of latite show near the top of the scarp, resting on white tuffaceous strata.   In the middle distance is a low basalt vent or dome from which Snake River basalt has issued and spread to the immediate foreground.   This vent is very near the base of the range and its position has very likely been determined by the faulting.   This part of the fault scarp has been little modified by landslides and its excellent stage of preservation argues for the recency of the fault movement.

A. ALBION OVERTHRUST

The inked line outlines the plane of the low-angled Albion overthrust. Above are the quartzitic and schist beds of the pre-Cambrian Harrison series and below the Pennsylvanian limestones and sandstones. This is the largest overthrust in the region and perhaps one of the greatest in the Rocky Mountain area.

B. OAKLEY FAULT SCARP

Young fault scarp in the Tertiary lavas near Oakley. The scarp rises about 500 feet and exposes four flows of latite and possibly five. The youthfulness of the scarp testifies to the recency of the Basin-Range normal or block faulting.

A. TUFF QUARRY

Part of quarry face seven miles southeast of Albion. Shows the bedding of the light colored pumiceous ash or tuff. Above may be seen the overlying flow of columnar quartz latite. Despite its apparent fragility the tuff serves well as a building stone.

B. BANK BUILDING IN ALBION CONSTRUCTED OF VOLCANIC TUFF

Shows the adaptability of the tuff as a building stone and its surprising resistance to crushing, which permits its use in two-story buildings. Darker colored blocks are of the yellowish-green tuff.

A.   LATITE QUARRY AT ALBION

Rock from the quarry has been used in the construction of the Albion State Normal School buildings.   It makes an attractive pinkish or pale reddish stone.   Some distance back may be seen the platy quartz latite.

B.   LATTER DAY SAINTS CHURCH IN OAKLEY CONSTRUCTED OF
QUARTZ LATITE BLOCKS

Shows how effectively the quartz latite may be used as a building stone.   It is possible to obtain well shaped blocks of this rock.   Cornerstones are of volcanic tuff.

www.GoldMiningBooks.com

Books On Mining

Visit: www.goldminingbooks.com to order your copies or ask your favorite book seller to offer them.

Mining Books by Kerby Jackson

Gold Dust: Stories From Oregon's Mining Years - Oregon mining historian and prospector, Kerby Jackson, brings you a treasure trove of seventeen stories on Southern Oregon's rich history of gold prospecting, the prospectors and their discoveries, and the breathtaking areas they settled in and made homes. 5" X 8", 98 ppgs. Retail Price: $11.99

The Golden Trail: More Stories From Oregon's Mining Years - In his follow-up to "Gold Dust: Stories of Oregon's Mining Years", this time around, Jackson brings us twelve tales from Oregon's Gold Rush, including the story about the first gold strike on Canyon Creek in Grant County, about the old timers who found gold by the pail full at the Victor Mine near Galice, how Iradel Bray discovered a rich ledge of gold on the Coquille River during the height of the Rogue River War, a tale of two elderly miners on the hunt for a lost mine in the Cascade Mountains, details about the discovery of the famous Armstrong Nugget and others. 5" X 8", 70 ppgs. Retail Price: $10.99

Oregon Mining Books

Geology and Mineral Resources of Josephine County, Oregon - Unavailable since the 1970's, this important publication was originally compiled by the Oregon Department of Geology and Mineral Industries and includes important details on the economic geology and mineral resources of this important mining area in South Western Oregon. Included are notes on the history, geology and development of important mines, as well as insights into the mining of gold, copper, nickel, limestone, chromium and other minerals found in large quantities in Josephine County, Oregon. 8.5" X 11", 54 ppgs. Retail Price: $9.99

Mines and Prospects of the Mount Reuben Mining District - Unavailable since 1947, this important publication was originally compiled by geologist Elton Youngberg of the Oregon Department of Geology and Mineral Industries and includes detailed descriptions, histories and the geology of the Mount Reuben Mining District in Josephine County, Oregon. Included are notes on the history, geology, development and assay statistics, as well as underground maps of all the major mines and prospects in the vicinity of this much neglected mining district. 8.5" X 11", 48 ppgs. Retail Price: $9.99

The Granite Mining District - Notes on the history, geology and development of important mines in the well known Granite Mining District which is located in Grant County, Oregon. Some of the mines discussed include the Ajax, Blue Ribbon, Buffalo, Continental, Cougar-Independence, Magnolia, New York, Standard and the Tillicum. Also included are many rare maps pertaining to the mines in the area. 8.5" X 11", 48 ppgs. Retail Price: $9.99

Ore Deposits of the Takilma and Waldo Mining Districts of Josephine County, Oregon - The Waldo and Takilma mining districts are most notable for the fact that the earliest large scale mining of placer gold and copper in Oregon took place in these two areas. Included are details about some of the earliest large gold mines in the state such as the Llano de Oro, High Gravel, Cameron, Platerica, Deep Gravel and others, as well as copper mines such as the famous Queen of Bronze mine, the Waldo, Lily and Cowboy mines. This volume also includes six maps and 20 original illustrations. 8.5" X 11", 74 ppgs. Retail Price: $9.99

Metal Mines of Douglas, Coos and Curry Counties, Oregon - Oregon mining historian Kerby Jackson introduces us to a classic work on Oregon's mining history in this important re-issue of Bulletin 14C Volume 1, otherwise known as the Douglas, Coos & Curry Counties, Oregon Metal Mines Handbook. Unavailable since 1940, this important publication was originally compiled by the Oregon Department of Geology and Mineral Industries includes detailed descriptions, histories and the geology of over 250 metallic mineral mines and prospects in this rugged area of South West Oregon. 8.5" X 11", 158 ppgs. Retail Price: $19.99

Metal Mines of Jackson County, Oregon - Unavailable since 1943, this important publication was originally compiled by the Oregon Department of Geology and Mineral Industries includes detailed descriptions, histories and the geology of over 450 metallic mineral mines and prospects in Jackson County, Oregon. Included are such famous gold mining areas as Gold Hill, Jacksonville, Sterling and the Upper Applegate. **8.5" X 11", 220 ppgs. Retail Price: $24.99**

Metal Mines of Josephine County, Oregon - Oregon mining historian Kerby Jackson introduces us to a classic work on Oregon's mining history in this important re-issue of Bulletin 14C, otherwise known as the Josephine County, Oregon Metal Mines Handbook. Unavailable since 1952, this important publication was originally compiled by the Oregon Department of Geology and Mineral Industries includes detailed descriptions, histories and the geology of over 500 metallic mineral mines and prospects in Josephine County, Oregon. **8.5" X 11", 250 ppgs. Retail Price: $24.99**

Metal Mines of North East Oregon - Oregon mining historian Kerby Jackson introduces us to a classic work on Oregon's mining history in this important re-issue of Bulletin 14A and 14B, otherwise known as the North East Oregon Metal Mines Handbook. Unavailable since 1941, this important publication was originally compiled by the Oregon Department of Geology and Mineral Industries and includes detailed descriptions, histories and the geology of over 750 metallic mineral mines and prospects in North Eastern Oregon. **8.5" X 11", 310 ppgs. Retail Price: $29.99**

Metal Mines of North West Oregon - Oregon mining historian Kerby Jackson introduces us to a classic work on Oregon's mining history in this important re-issue of Bulletin 14D, otherwise known as the North West Oregon Metal Mines Handbook. Unavailable since 1951, this important publication was originally compiled by the Oregon Department of Geology and Mineral Industries and includes detailed descriptions, histories and the geology of over 250 metallic mineral mines and prospects in North Western Oregon. **8.5" X 11", 182 ppgs. Retail Price: $19.99**

Mines and Prospects of Oregon - Mining historian Kerby Jackson introduces us to a classic mining work by the Oregon Bureau of Mines in this important re-issue of The Handbook of Mines and Prospects of Oregon. Unavailable since 1916, this publication includes important insights into hundreds of gold, silver, copper, coal, limestone and other mines that operated in the State of Oregon around the turn of the 19th Century. Included are not only geological details on early mines throughout Oregon, but also insights into their history, production, locations and in some cases, also included are rare maps of their underground workings. **8.5" X 11", 314 ppgs. Retail Price: $24.99**

Lode Gold of the Klamath Mountains of Northern California and South West Oregon
(See California Mining Books)

Mineral Resources of South West Oregon - Unavailable since 1914, this publication includes important insights into dozens of mines that once operated in South West Oregon, including the famous gold fields of Josephine and Jackson Counties, as well as the Coal Mines of Coos County. Included are not only geological details on early mines throughout South West Oregon, but also insights into their history, production and locations. **8.5" X 11", 154 ppgs. Retail Price: $11.99**

Chromite Mining in The Klamath Mountains of California and Oregon
(See California Mining Books)

Southern Oregon Mineral Wealth - Unavailable since 1904, this rare publication provides a unique snapshot into the mines that were operating in the area at the time. Included are not only geological details on early mines throughout South West Oregon, but also insights into their history, production and locations. Some of the mining areas include Grave Creek, Greenback, Wolf Creek, Jump Off Joe Creek, Granite Hill, Galice, Mount Reuben, Gold Hill, Galls Creek, Kane Creek, Sardine Creek, Birdseye Creek, Evans Creek, Foots Creek, Jacksonville, Ashland, the Applegate River, Waldo, Kerby and the Illinois River, Althouse and Sucker Creek, as well as insights into local copper mining and other topics. **8.5" X 11", 64 ppgs. Retail Price: $8.99**

Geology and Ore Deposits of the Takilma and Waldo Mining Districts - Unavailable since the 1933, this publication was originally compiled by the United States Geological Survey and includes details on gold and copper mining in the Takilma and Waldo Districts of Josephine County, Oregon. The Waldo and Takilma mining districts are most notable for the fact that the earliest large scale mining of placer gold and copper in Oregon took place in these two areas. Included in this report are details about some of the earliest large gold mines in the state such as the Llano de Oro, High Gravel, Cameron, Platerica, Deep Gravel and others, as well as copper mines such as the famous Queen of Bronze mine, the Waldo, Lily and Cowboy mines. In addition to geological examinations, insights are also provided into the production, day to day operations and early histories of these mines, as well as calculations of known mineral reserves in the area. This volume also includes six maps and 20 original illustrations. **8.5" X 11", 74 ppgs. Retail Price: $9.99**

Gold Mines of Oregon - Oregon mining historian Kerby Jackson introduces us to a classic work on Oregon's mining history in this important re-issue of Bulletin 61, otherwise known as "Gold and Silver In Oregon". Unavailable since 1968, this important publication was originally compiled by geologists Howard C. Brooks and Len Ramp of the Oregon Department of Geology and Mineral Industries and includes detailed descriptions, histories and the geology of over 450 gold mines Oregon. Included are notes on the history, geology and gold production statistics of all the major mining areas in Oregon including the Klamath Mountains, the Blue Mountains and the North Cascades. While gold is where you find it, as every miner knows, the path to success is to prospect for gold where it was previously found. 8.5" X 11", 344 ppgs. **Retail Price: $24.99**

Mines and Mineral Resources of Curry County Oregon - Originally published in 1916, this important publication on Oregon Mining has not been available for nearly a century. Included are rare insights into the history, production and locations of dozens of gold mines in Curry County, Oregon, as well as detailed information on important Oregon mining districts in that area such as those at Agness, Bald Face Creek, Mule Creek, Boulder Creek, China Diggings, Collier Creek, Elk River, Gold Beach, Rock Creek, Sixes River and elsewhere. Particular attention is especially paid to the famous beach gold deposits of this portion of the Oregon Coast. 8.5" X 11", 140 ppgs. **Retail Price: $11.99**

Chromite Mining in South West Oregon - Originally published in 1961, this important publication on Oregon Mining has not been available for nearly a century. Included are rare insights into the history, production and locations of nearly 300 chromite mines in South Western Oregon. 8.5" X 11", 184 ppgs. **Retail Price: $14.99**

Mineral Resources of Douglas County Oregon - Originally published in 1972, this important publication on Oregon Mining has not been available for nearly forty years. Included are rare insights into the geology, history, production and locations of numerous gold mines and other mining properties in Douglas County, Oregon. 8.5" X 11", 124 ppgs. **Retail Price: $11.99**

Mineral Resources of Coos County Oregon - Originally published in 1972, this important publication on Oregon Mining has not been available for nearly forty years. Included are rare insights into the geology, history, production and locations of numerous gold mines and other mining properties in Coos County, Oregon. 8.5" X 11", 100 ppgs. **Retail Price: $11.99**

Mineral Resources of Lane County Oregon - Originally published in 1938, this important publication on Oregon Mining has not been available for nearly seventy five years. Included are extremely rare insights into the geology and mines of Lane County, Oregon, in particular in the Bohemia, Blue River, Oakridge, Black Butte and Winberry Mining Districts. 8.5" X 11", 82 ppgs. **Retail Price: $9.99**

Mineral Resources of the Upper Chetco River of Oregon: Including the Kalmiopsis Wilderness - Originally published in 1975, this important publication on Oregon Mining has not been available for nearly forty years. Withdrawn under the 1872 Mining Act since 1984, real insight into the minerals resources and mines of the Upper Chetco River has long been unavailable due to the remoteness of the area. Despite this, the decades of battle between property owners and environmental extremists over the last private mining inholding in the area has continued to pique the interest of those interested in mining and other forms of natural resource use. Gold mining began in the area in the 1850's and has a rich history in this geographic area, even if the facts surrounding it are little known. Included are twenty two rare photographs, as well as insights into the Becca and Morning Mine, the Emmly Mine (also known as Emily Camp), the Frazier Mine, the Golden Dream or Higgins Mine, Hustis Mine, Peck Mine and others. 8.5" X 11", 64 ppgs. **Retail Price: $8.99**

Gold Dredging in Oregon - Originally published in 1939, this important publication on Oregon Mining has not been available for nearly seventy five years. Included are extremely rare insights into the history and day to day operations of the dragline and bucketline gold dredges that once worked the placer gold fields of South West and North East Oregon in decades gone by. Also included are details into the areas that were worked by gold dredges in Josephine, Jackson, Baker and Grant counties, as well as the economic factors that impacted this mining method. This volume also offers a unique look into the values of river bottom land in relation to both farming and mining, in how farm lands were mined, re-soiled and reclamated after the dredges worked them. Featured are hard to find maps of the gold dredge fields, as well as rare photographs from a bygone era. 8.5" X 11", 86 ppgs. **Retail Price: $8.99**

Quick Silver Mining in Oregon - Originally published in 1963, this important publication on Oregon Mining has not been available for over fifty years. This publication includes details into the history and production of Elemental Mercury or Quicksilver in the State of Oregon. 8.5" X 11", 238 ppgs. **Retail Price: $15.99**

Mines of the Greenhorn Mining District of Grant County Oregon - Originally published in 1948, this important publication on Oregon Mining has not been available for over sixty five years. In this publication are rare insights into the mines of the famous Greenhorn Mining District of Grant County, Oregon, especially the famous Morning Mine. Also included are details on the Tempest, Tiger, Bi-Metallic, Windsor, Psyche, Big Johnny, Snow Creek, Banzette and Paramount Mines, as well as prospects in the vicinities in the famous mining areas of Mormon Basin, Vinegar Basin and Desolation Creek. Included are hard to find mine maps and dozens of rare photographs from the bygone era of Grant County's rich mining history. 8.5" X 11", 72 ppgs. **Retail Price: $9.99**

Geology of the Wallowa Mountains of Oregon: Part I (Volume 1) - Originally published in 1938, this important publication on Oregon Mining has not been available for nearly seventy five years. Included are details on the geology of this unique portion of North Eastern Oregon. This is the first part of a two book series on the area. Accompanying the text are rare photographs and historic maps. **8.5" X 11", 92 ppgs. Retail Price: $9.99**

Geology of the Wallowa Mountains of Oregon: Part II (Volume 2) - Originally published in 1938, this important publication on Oregon Mining has not been available for nearly seventy five years. Included are details on the geology of this unique portion of North Eastern Oregon. This is the first part of a two book series on the area. Accompanying the text are rare photographs and historic maps. **8.5" X 11", 94 ppgs. Retail Price: $9.99**

Field Identification of Minerals For Oregon Prospectors - Originally published in 1940, this important publication on Oregon Mining has not been available for nearly seventy five years. Included in this volume is an easy system for testing and identifying a wide range of minerals that might be found by prospectors, geologists and rockhounds in the State of Oregon, as well as in other locales. Topics include how to put together your own field testing kit and how to conduct rudimentary tests in the field. This volume is written in a clear and concise way to make it useful even for beginners. **8.5" X 11", 158 ppgs. Retail Price: $14.99**

The Bohemia Mining District of Oregon - Originally published in 1900, this important publication on Oregon Mining has not been available for over a century. Included in this volume are important insights into the famous Bohemia Mining District of Oregon, including the histories and locations of important gold mines in the area such as the Ophir Mine, Clarence, Acturas, Peek-a-boo, White Swan, Combination Mine, the Musick Mine, The California, White Ghost, The Mystery, Wall Street, Vesuvius, Story, Lizzie Bullock, Delta, Elsie Dora, Golden Slipper, Broadway, Champion Mine, Knott, Noonday, Helena, White Wings, Riverside and others. Also included are notes on the nearby Blue River Mining District. **8.5" X 11", 58 ppgs. Retail Price: $9.99**

The Gold Fields of Eastern Oregon - Unavailable since 1900, this publication was originally compiled by the Baker City Chamber of Commerce Offering important insights into the gold mining history of Eastern Oregon, "The Gold Fields of Eastern Oregon" sheds a rare light on many of the gold mines that were operating at the turn of the 19th Century in Baker County and Grant County in North Eastern Oregon. Some of the areas featured include the Cable Cove District, Baisely-Elhorn, Granite, Red Boy, Bonanza, Susanville, Sparta, Virtue, Vaughn, Sumpter, Burnt River, Rye Valley and other mining districts. Included is basic information on not only many gold mines that are well known to those interested in Eastern Oregon mining history, but also many mines and prospects which have been mostly lost to the passage of time. Accompanying are numerous rare photos **8.5" X 11", 78 ppgs. Retail Price: $10.99**

Gold Mining in Eastern Oregon - Originally published in 1938, this important publication on Oregon Mining has not been available for over a century. Included in this volume are important insights into the famous mining districts of Eastern Oregon during the late 1930's. Particular attention is given to those gold mines with milling and concentrating facilities in the Greenhorn, Red Boy, Alamo, Bonanza, Granite, Cable Cove, Cracker Creek, Virtue, Keating, Medical Springs, Sanger, Sparta, Chicken Creek, Mormon Basin, Connor Creek, Cornucopia and the Bull Run Mining Districts. Some of the mines featured include the Ben Harrison, North Pole-Columbia, Highland Maxwell, Baisley-Elkhorn, White Swan, Balm Creek, Twin Baby, Gem of Sparta, New Deal, Gleason, Gifford-Johnson, Cornucopia, Record, Bull Run, Orion and others. Of particular interest are the mill flow sheets and descriptions of milling operations of these mines. **8.5" X 11", 68 ppgs. Retail Price: $8.99**

The Gold Belt of the Blue Mountains of Oregon - Originally published in 1901, this important publication on Oregon Mining has not been available for over a century. Included in this volume are rare insights into the gold deposits of the Blue Mountains of North East Oregon, including the history of their early discovery and early production. Extensive details are offered on this important mining area's mineralogy and economic geology, as well as insights into nearby gold placers, silver deposits and copper deposits. Featured are the Elkhorn and Rock Creek mining districts, the Pocahontas district, Auburn and Minersville districts, Sumpter and Cracker Creek, Cable Cove, the Camp Carson district, Granite, Alamo, Greenhorn, Robinsonville, the Upper Burnt River Valley and Bonanza districts, Susanville, Quartzburg, Canyon Creek, Virtue, the Copper Butte district, the North Powder River, Sparta, Eagle Creek, Cornucopia, Pine Creek, Lower Powder River, the Upper Snake River Canyon, Rye Valley, Lower Burnt River Valley, Mormon Basin, the Malheur and Clarks Creek districts, Sutton Creek and others. Of particular interest are important details on numerous gold mines and prospects in these mining districts, including their locations, histories, geology and other important information, as well as information on silver, copper and fire opal deposits. **8.5" X 11", 250 ppgs. Retail Price: $24.99**

Mining in the Cascades Range of Oregon - Originally published in 1938, this important publication on Oregon Mining has not been available for over seventy five years. Included in this volume are rare insights into the gold mines and other types of metal mines in the Cascades Mountain Range of Oregon. Some of the important mining areas covered include the famous Bohemia Mining District, the North Santiam Mining District, Quartzville Mining District, Blue River Mining District, Fall Creek Mining District, Oakridge District, Zinc District, Buzzard-Al Sarena District, Grand Cove, Climax District and Barron Mining District. Of particular interest are important details on over 100 mines and prospects in these mining districts, including their locations, histories, geology and other important information. 8.5" X 11", 170 ppgs. Retail Price: $14.99

Beach Gold Placers of the Oregon Coast - Originally published in 1934, this important publication on Oregon Mining has not been available for over 80 years. Included in this volume are rare insights into the beach gold deposits of the State of Oregon, including their locations, occurance, composition and geology. Of particular interest is information on placer platinum in Oregon's rich beach deposits. Also included are the locations and other information on some famous Oregon beach mines, including the Pioneer, Eagle, Chickamin, Iowa and beach placer mines north of the mouth of the Rogue River. 8.5" X 11", 60 ppgs. Retail Price: $8.99

Mineralogical Composition of the Sands of the Oregon Coast: From Coos Bay to the Columbia - Published in 1945, he text features hard to find information on the composition of the gold bearing black sands of the South West Oregon Coast, offering a unique insight to prospectors in search of Oregon's legendary beach gold. 104 ppgs, $9.99

Manganese Mining in Oregon - First released in 1942 and now out of print, this special reprint edition of "Manganese in Oregon" was originally published by the Oregon Department of Geology and Mineral Industries. The text features hard to find information on the mining of Manganese in Oregon, including details and maps of Oregon manganese mines and prospects. 108 ppgs, 9.99

Medford Oregon As A Mining Center - Written in 1912, this hard to find publication includes valuable insights into the mining history of South West Oregon. This small book contains interesting information on the gold, copper and mining industry in Southern Oregon as it existed just prior to World War One, shedding light on some of the important mines in the area. Included are rare photographs and vintage advertising of the day. 80 ppgs, 9.99

Mineral Resources of Curry County Oregon - First released in 1977 and now out of print, this special reprint edition of "Geology, Mineral Resources and Rock Materials of Curry County, Oregon" was originally published in cooperation of Curry County, Oregon and the Oregon Department of Geology and Mineral Industries. The text features hard to find information on not only the mining of gold and other metals in Curry County, but also aggregate mining in the area. 102 ppgs, 11.99

Origin of the Gold Bearing Black Sands of the Coast of South West Oregon - First released in 1943 and now out of print, this special reprint edition of "The Origin of the Black Sands of the South West Oregon Coast" was originally published by the Oregon Department of Geology and Mineral Industries. The text features hard to find information on the origin of the gold bearing black sands of the South West Oregon Coast, offering a unique insight to prospectors in search of Oregon's legendary beach gold. 52 ppgs, 8.99

South West Oregon Mining - Leading mining historian Kerby Jackson introduces us to six classic small mining publications on the Gold Mining Industry in Southern Oregon. This small book consists of a compilation of USGS J.S. Diller's "Mines of the Riddles Quadrangle", "The Rogue River Valley Coal Fields" and "Mineral Resources of the Grants Pass Quadrangle", the Grants Pass Commercial Club's rare publication "Mining in Josephine County, Oregon" and the USGS publication "The Distribution of Placer Gold in the Sixes River, South West Oregon". Also included is F.W. Libbey's legendary article on the Southern Oregon Mining Industry, "Lest We Forget", which appeared in the publication of the Oregon State Department of Geology and Mineral Industries in the early 1960's. This compilation offers a unique perspective on mining in South West Oregon and includes considerable information on mines in Josephine, Jackson and Coos Counties. 142 ppgs, 14.99

Geology and Mineral Resources of the Gasquet Quadrangle of California-Oregon - First published in 1953, it has been unavailable for over a century and sheds important light on the geological features and mineral resources of this portion of Northern California and Southern Oregon. 80 ppgs, 9.99

Idaho Mining Books

Gold in Idaho - Unavailable since the 1940's, this publication was originally compiled by the Idaho Bureau of Mines and includes details on gold mining in Idaho. Included is not only raw data on gold production in Idaho, but also valuable insight into where gold may be found in Idaho, as well as practical information on the gold bearing rocks and other geological features that will assist those looking for placer and lode gold in the State of Idaho. This volume also includes thirteen gold maps that greatly enhance the practical usability of the information contained in this small book detailing where to find gold in Idaho. **8.5" X 11", 72 ppgs. Retail Price: $9.99**

Geology of the Couer D'Alene Mining District of Idaho - Unavailable since 1961, this publication was originally compiled by the Idaho Bureau of Mines and Geology and includes details on the mining of gold, silver and other minerals in the famous Coeur D'Alene Mining District in Northern Idaho. Included are details on the early history of the Coeur D'Alene Mining District, local tectonic settings, ore deposit features, information on the mineral belts of the Osburn Fault, as well as detailed information on the famous Bunker Hill Mine, the Dayrock Mine, Galena Mine, Lucky Friday Mine and the infamous Sunshine Mine. This volume also includes sixteen hard to find maps. **8.5" X 11", 70 ppgs. Retail Price: $9.99**

The Gold Camps and Silver Cities of Idaho - Originally published in 1963, this important publication on Idaho Mining has not been available for nearly fifty years. Included are rare insights into the history of Idaho's Gold Rush, as well as the mad craze for silver in the Idaho Panhandle. Documented in fine detail are the early mining excitements at Boise Basin, at South Boise, in the Owyhees, at Deadwood, Long Valley, Stanley Basin and Robinson Bar, at Atlanta, on the famous Boise River, Volcano, Little Smokey, Banner, Boise Ridge, Hailey, Leesburg, Lemhi, Pearl, at South Mountain, Shoup and Ulysses, Yellow Jacket and Loon Creek. The story follows with the appearance of Chinese miners at the new mining camps on the Snake River, Black Pine, Yankee Fork, Bay Horse, Clayton, Heath, Seven Devils, Gibbonsville, Vienna and Sawtooth City. Also included are special sections on the Idaho Lead and Silver mines of the late 1800's, as well as the mining discoveries of the early 1900's that paved the way for Idaho's modern mining and mineral industry. Lavishly illustrated with rare historic photos, this volume provides a one of a kind documentary into Idaho's mining history that is sure to be enjoyed by not only modern miners and prospectors who still scour the hills in search of nature's treasures, but also those enjoy history and tromping through overgrown ghost towns and long abandoned mining camps. **8.5" X 11", 186 ppgs. Retail Price: $14.99**

Ore Deposits and Mining in North Western Custer County Idaho - Unavailable since 1913, this important publication was originally published by the Us Department of the Interior and has been unavailable for a century. Included are fine details on the geology, geography, gold placers and gold and silver bearing quartz veins of the mining region of North West Custer County, Idaho. Of particular interest is a rare look at the mines and prospects of the region, including those such as the Ramshorn Mine, SkyLark, Riverview, Excelsior, Beardsley, Pacific, Hoosier, Silver Brick, Forest Rose and dozens of others in the Bay Horse Mining District. Also covered are the mines of the Yankee Fork District such as the Lucky Boy, Badger, Black, Enterprise, Charles Dickens, Morrison, Golden Sunbeam, Montana, Golden Gate and others, as well as those in the Loon Mining District. **8.5" X 11", 126 ppgs. Retail Price: $12.99**

Gold Rush To Idaho - Unavailable since 1963, this important publication was originally published by the Idaho Bureau of Mines and has been unavailable for 50 years. "Gold Rush To Idaho" revisits the earliest years of the discovery of gold in Idaho Territory and introduces us to the conditions that the pioneer gold seekers met when they blazed a trail through the wilderness of Idaho's mountains and discovered the precious yellow metal at Oro Fino and Pierce. Subsequent rushes followed at places like Elk City, Newsome, Clearwater Station, Florence, Warrens and elsewhere. Of particular interest is a rare look at the hardships that the first miners in Idaho met with during their day to day existences and their attempts to bring law and order to their mining camps. **8.5" X 11", 88 ppgs. Retail Price: $9.99**

The Geology and Mines of Northern Idaho and North Western Montana - Unavailable since 1909, this important publication was originally published by the Us Department of the Interior and has been unavailable for a century. Included are fine details on the geology and geography of the mining regions of Northern Idaho and North Western Montana. Of particular interest is a rare look at the mines and prospects of the region, including those in the Pine Creek Mining District, Lake Pend Oreille district, Troy Mining District, Sylvanite District, Cabinet Mining District, Prospect Mining District and the Missoula Valley. Some of the mines featured include the Iron Mountain, Silver Butte, Snowshoe, Grouse Mountain Mine and others. **8.5" X 11", 142 ppgs. Retail Price: $12.99**

Mining in the Alturas Quadrangle of Blaine County Idaho - Unavailable since 1922, this important publication was originally published by the Idaho Bureau of Mines and has been unavailable for ninety years. Topics include the geology, rock formations and the formation of ore deposits in this important mining area of Idaho. Of particular focus is information on the local geology, quartz veins and ore deposits of this portion of Idaho. Included are hard to find details, including the descriptions and locations of numerous gold and silver mines in the area including the Silver King, Pilgrim, Columbia, Lone Jack, Sunbeam, Pride of the West, Lucky Boy, Scotia, Atlanta, Beaver-Bidwell and others mines and prospects. **8.5" X 11", 56 ppgs. Retail Price: $8.99**

Mining in Lemhi County Idaho - Originally published in 1913, this important book on Idaho Mining has not been available to miners for over a century. Included are rare insights into hundreds of gold, silver, copper and other mines in this famous Idaho mining area. Details include the locations, geology, history, production and other facts of the mines of this region, not only gold and silver hardrock mines, but also gold placer mines, lead-silver deposits, copper mines, cobalt-nickel deposits, tungsten and tin mines . It is lavishly illustrated with hard to find photos of the period and rare mining maps. Some of the vicinities featured include the Nicholia Mining District, Spring Mountain District, Texas District, Blue Wing District, Junction District, McDevitt District, Pratt Creek, Eldorado District, Kirtley Creek, Carmen Creek, Gibbonsville, Indian Creek, Mineral Hill District, Mackinaw, Eureka District, Blackbird District, YellowJacket District, Gravel Range District, Junction District, Parker Mountain and other mining districts. 8.5" X 11", 226 ppgs. Retail Price: $19.99

Mining in Shoshone County Idaho - First published in 1923, it has been unavailable for over a century and sheds important light on the mining history of Shoshone County, Idaho. Some of the topics include the history of mining in Shoshone County, a look at the local geology and ore characteristics of lead-silver deposits, zinc deposits, copper, antimony, gold and other minerals. Also included are insights into the history, production, characteristics and locations of numerous mines in the area. 198 ppgs, 15.99

Utah Mining Books

Fluorite in Utah - Unavailable since 1954, this publication was originally compiled by the USGS, State of Utah and U.S. Atomic Energy Commission and details the mining of fluorspar, also known as fluorite in the State of Utah. Included are details on the geology and history of fluorspar (fluorite) mining in Utah, including details on where this unique gem mineral may be found in the State of Utah. 8.5" X 11", 60 ppgs. Retail Price: $8.99

The Gold Hill Mining District of Utah - First published in 1935, it has been unavailable since those days and sheds important light on the mines, history and geology of Utah's Gold Hill Mining District. Included are rare insights into this important mining area, including the locations, histories and details of numerous mines. This volume is well illustrated with geological diagrams, as well as hard to find maps of some of the most important mines in this district. 202 ppgs., 19.99

The Mines, Miners and Minerals of Utah - First published in 1896, it has been unavailable since those days and sheds important light on the early mines and miners of Pioneer Utah, as well as the minerals which they won from the earth by laborious hard physical labor and sheer determination. Included are rare insights into the early mining history of Utah, as well details on hundreds of gold, silver and copper mines. 376 ppgs., 24.99

California Mining Books

The Tertiary Gravels of the Sierra Nevada of California - Mining historian Kerby Jackson introduces us to a classic mining work by Waldemar Lindgren in this important re-issue of The Tertiary Gravels of the Sierra Nevada of California. Unavailable since 1911, this publication includes details on the gold bearing ancient river channels of the famous Sierra Nevada region of California. 8.5" X 11", 282 ppgs. Retail Price: $19.99

The Mother Lode Mining Region of California - Unavailable since 1900, this publication includes details on the gold mines of California's famous Mother Lode gold mining area. Included are details on the geology, history and important gold mines of the region, as well as insights into historic mining methods, mine timbering, mining machinery, mining bell signals and other details on how these mines operated. Also included are insights into the gold mines of the California Mother Lode that were in operation during the first sixty years of California's mining history. 8.5" X 11", 176 ppgs. Retail Price: $14.99

Lode Gold of the Klamath Mountains of Northern California and South West Oregon - Unavailable since 1971, this publication was originally compiled by Preston E. Hotz and includes details on the lode mining districts of Oregon and California's Klamath Mountains. Included are details on the geology, history and important lode mines of the French Gulch, Deadwood, Whiskeytown, Shasta, Redding, Muletown, South Fork, Old Diggings, Dog Creek (Delta), Bully Choop (Indian Creek), Harrison Gulch, Hayfork, Minersville, Trinity Center, Canyon Creek, East Fork, New River, Denny, Liberty (Black Bear), Cecilville, Callahan, Yreka, Fort Jones and Happy Camp mining districts in California, as well as the Ashland, Rogue River, Applegate, Illinois River, Takilma, Greenback, Galice, Silver Peak, Myrtle Creek and Mule Creek districts of South Western Oregon. Also included are insights into the mineralization and other characteristics of this important mining region. 8.5" X 11", 100 ppgs. Retail Price: $10.99

Mines and Mineral Resources of Shasta County, Siskiyou County, Trinity County: California - Unavailable since 1915, this publication was originally compiled by the California State Mining Bureau and includes details on the gold mines of this area of Northern California. Also included are insights into the mineralization and other characteristics of this important mining region, as well as the location of historic gold mines. 8.5" X 11", 204 ppgs. Retail Price: $19.99

Geology of the Yreka Quadrangle, Siskiyou County, California - Unavailable since 1977, this publication was originally compiled by Preston E. Hotz and includes details on the geology of the Yreka Quadrangle of Siskiyou County, California. Also included are insights into the mineralization and other characteristics of this important mining region. **8.5" X 11", 78 ppgs. Retail Price: $7.99**

Mines of San Diego and Imperial Counties, California - Originally published in 1914, this important publication on California Mining has not been available for a century. This publication includes important information on the early gold mines of San Diego and Imperial County, which were some of the first gold fields mined in California by early Spanish and Mexican miners before the 49ers came on the scene. Included are not only details on early mining methods in the area, production statistics and geological information, but also the location of the early gold mines that helped make California "The Golden State". Also included are details on the mining of other minerals such as silver, lead, zinc, manganese, tungsten, vanadium, asbestos, barite, borax, cement, clay, dolomite, fluospar, gem stones, graphite, marble, salines, petroleum, stronium, talc and others. **8.5" X 11", 116 ppgs. Retail Price: $12.99**

Mines of Sierra County, California - Unavailable since 1920, this publication was originally compiled by the California State Mining Bureau and includes details on the gold mines of Sierra County, California. Also included are insights into the mineralization and other characteristics of this important mining region, as well as the location of historic gold mines. **8.5" X 11", 156 ppgs. Retail Price: $19.99**

Mines of Plumas County, California - Unavailable since 1918, this publication was originally compiled by the California State Mining Bureau and includes details on the gold mines of Plumas County, California. Also included are insights into the mineralization and other characteristics of this important mining region, as well as the location of historic gold mines. **8.5" X 11", 200 ppgs. Retail Price: $19.99**

Mines of El Dorado, Placer, Sacramento and Yuba Counties, California - Originally published in 1917, this important publication on California Mining has not been available for nearly a century. This publication includes important information on the early gold mines of El Dorado County, Placer County, Sacramento County and Yuba County, which were some of the first gold fields mined by the Forty-Niners during the California Gold Rush. Included are not only details on early mining methods in the area, production statistics and geological information, but also the location of the early gold mines that helped make California "The Golden State". Also included are insights into the early mining of chrome, copper and other minerals in this important mining area. **8.5" X 11", 204 ppgs. Retail Price: $19.99**

Mines of Los Angeles, Orange and Riverside Counties, California - Originally published in 1917, this important publication on California Mining has not been available for nearly a century. This publication includes important information on the early gold mines of Los Angeles County, Orange County and Riverside County, which were some of the first gold fields mined in California by early Spanish and Mexican miners before the 49ers came on the scene. Included are not only details on early mining methods in the area, production statistics and geological information, but also the location of the early gold mines that helped make California "The Golden State". **8.5" X 11", 146 ppgs. Retail Price: $12.99**

Mines of San Bernadino and Tulare Counties, California - Originally published in 1917, this important publication on California Mining has not been available for nearly a century. This publication includes important information on the early gold mines of San Bernadino and Tulare County, which were some of the first gold fields mined in California by early Spanish and Mexican miners before the 49ers came on the scene. Included are not only details on early mining methods in the area, production statistics and geological information, but also the location of the early gold mines that helped make California "The Golden State". Also included are details on the mining of other minerals such as copper, iron, lead, zinc, manganese, tungsten, vanadium, asbestos, barite, borax, cement, clay, dolomite, fluospar, gem stones, graphite, marble, salines, petroleum, stronium, talc and others. **8.5" X 11", 200 ppgs. Retail Price: $19.99**

Chromite Mining in The Klamath Mountains of California and Oregon - Unavailable since 1919, this publication was originally compiled by J.S. Diller of the United States Department of Geological Survey and includes details on the chromite mines of this area of Northern California and Southern Oregon. Also included are insights into the mineralization and other characteristics of this important mining region, as well as the location of historic mines. Also included are insights into chromite mining in Eastern Oregon and Montana. **8.5" X 11", 98 ppgs. Retail Price: $9.99**

Mines and Mining in Amador, Calaveras and Tuolumne Counties, California - Unavailable since 1915, this publication was originally compiled by William Tucker and includes details on the mines and mineral resources of this important California mining area. Included are details on the geology, history and important gold mines of the region, as well as insights into other local mineral resources such as asbestos, clay, copper, talc, limestone and others. Also included are insights into the mineralization and other characteristics of this important portion of California's Mother Lode mining region. **8.5" X 11", 198 ppgs. Retail Price: $14.99**

The Cerro Gordo Mining District of Inyo County California - Unavailable since 1963, this publication was originally compiled by the United States Department of Interior. Included are insights into the mineralization and other characteristics of this important mining region of Southern California. Topics include the mining of gold and silver in this important mining district in Inyo County, California, including details on the history, production and locations of the Cerro Gordo Mine, the Morning Star Mine, Estelle Tunnel, Charles Lease Tunnel, Ignacio, Hart, Crosscut Tunnel, Sunset, Upper Newtown, Newtown, Ella, Perseverance, Newsboy, Belmont and other silver and gold mines in the Cerro Gordo Mining District. This volume also includes important insights into the fossil record, geologic formations, faults and other aspects of economic geology in this California mining district. **8.5" X 11", 104 ppgs. Retail Price: $10.99**

Mining in Butte, Lassen, Modoc, Sutter and Tehama Counties of California - Unavailable since 1917, this publication was originally compiled by the United States Department of Interior. Included are insights into the mineralization and other characteristics of this important mining region of California. Topics include the mining of asbestos, chromite, gold, diamonds and manganese in Butte County, the mining of gold and copper in the Hayden Hill and Diamond Mountain mining districts of Lassen County, the mining of coal, salt, copper and gold in the High Grade and Winters mining districts of Modoc County, gold mining in Sutter County and the mining of gold, chromite, manganese and copper in Tehama County. This volume also includes the production records and locations of numerous mines in this important mining region. **8.5" X 11", 114 ppgs. Retail Price: $11.99**

Mines of Trinity County California - Originally published in 1965, this important publication on California Mining has not been available for nearly fifty years. This publication includes important information on mines and mining in Trinity County, California, as well insights into the mineralization and geology of this important mining area in Northern California. Included are extensive details on hardrock and placer gold mines and prospects, including charts showing the locations of these historic mines.. **8.5" X 11", 144 ppgs. Retail Price: $12.99**

Mines of Kern County California - Originally published in 1962, this important publication on California Mining has not been available for nearly fifty years. This publication includes important information on mines and mining in Kern County, California, as well insights into the mineralization and geology of this important mining area in California. Included are extensive details on hardrock and placer gold mines and prospects, including charts showing the locations of these historic mines. **8.5" X 11", 398 ppgs. Retail Price: $24.99**

Mines of Calaveras County California - Originally published in 1962, this important publication on California Mining has not been available for nearly fifty years. This publication includes important information on mines and mining in Calaveras County, California, as well insights into the mineralization and geology of this important mining area in Northern California. Included are extensive details on hardrock and placer gold mines and prospects, including charts showing the locations of these historic mines. **8.5" X 11", 236 ppgs. Retail Price: $19.99**

Lode Gold Mining in Grass Valley California - Unavailable since 1940, this publication was originally compiled by the United States Department of Interior. Included are insights into the gold mineralization and other characteristics of this important mining region of Nevada County, California. This volume also includes important insights into the geologic formations, faults and other aspects of economic geology in this California mining district. Of particular interest are the fine details on many hardrock gold mines in the area, including their locations, histories, development and mineralization. Some of the mines featured include the Gold Hill Mine, Massachusetts Hill, Boundary, Peabody, Golden Center, North Star, Omaha, Lone Jack, Homeward Bound, Hartery, Wisconsin, Allison Ranch, Phoenix, Kate Hayes, W.Y.O.D., Empire, Rich Hill, Daisy Hill, Orleans, Sultana, Centennial, Conlin, Ben Franklin, Crown Point and many others. **8.5" X 11", 148 ppgs. Retail Price: $12.99**

Lode Mining in the Alleghany District of Sierra County California - Unavailable since 1913, this publication was originally compiled by the United States Department of Interior. Included are insights into the mineralization and other characteristics of this important mining region of Sierra County. Included are details on the history, production and locations of numerous hardrock gold mines in this famous California area, including the Tightner Mine, Minnie D., Osceola, Eldorado, Twenty One, Sherman, Kenton, Oriental, Rainbow, Plumbago, Irelan, Gold Canyon, North Fork, Federal, Kate Hardy and others. This volume also includes important insights into the fossil record, geologic formations, faults and other aspects of economic geology in this California mining district. **8.5" X 11", 48 ppgs. Retail Price: $7.99**

Six Months In The Gold Mines During The California Gold Rush - Unavailable since 1850, this important work is a first hand account of one "49'ers" personal experience during the great California Gold Rush, shedding important light on one of the most exciting periods in the history of not only California, but also the world. Compiled from journals written between 1847 and 1849 by E. Gould Buffum, a native of New York, "Six Months In The Gold Mines During The California Gold Rush" offers a rare look into the day to day lives of the people who came to California to work in her gold mines when the state was still a great frontier. **8.5" X 11", 290 ppgs. Retail Price: $19.99**

<u>Quartz Mines of the Grass Valley Mining District of California</u> - Unavailable since 1867, this important publication has not been available since those days. This rare publication offers a short dissertation on the early hardrock mines in this important mining district in the California Mother Lode region between the 1850's and 1860's. Also included are hard to find details on the mineralization and locations of these mines, as well as how they were operated in those day. 8.5" X 11", 44 ppgs. Retail Price: $8.99

<u>Gold Rush on the Feather River</u> - First published in 1924, this short publication by G.C. Mansfield sheds important light on the early history of gold mining on the Feather River. Included are rare insights into the first decade of gold mining and the early mining camps of the Feather River during the 1850's. 64 ppgs., 9.99

<u>The Bodie Mining District of California</u> - First published in 1986, it has been unavailable since those days and sheds important light on this famous mining area. Included are the history, characteristics and locations of numerous old mines around the ghost town of Bodie. 64 ppgs, 8.99

<u>Geology and Mineral Resources of the Gasquet Quadrangle of California-Oregon</u> - First published in 1953, it has been unavailable for over a century and sheds important light on the geological features and mineral resources of this portion of Northern California and Southern Oregon. 80 ppgs, 9.99

Alaska Mining Books

<u>Ore Deposits of the Willow Creek Mining District, Alaska</u> - Unavailable since 1954, this hard to find publication includes valuable insights into the Willow Creek Mining District near Hatcher Pass in Alaska. The publication includes insights into the history, geology and locations of the well known mines in the area, including the Gold Cord, Independence, Fern, Mabel, Lonesome, Snowbird, Schroff-O'Neil, High Grade, Marion Twin, Thorpe, Webfoot, Kelly-Willow, Lane, Holland and others. 8.5" X 11", 96 ppgs. Retail Price: $9.99

<u>The Juneau Gold Belt of Alaska</u> - Unavailable since 1906, this hard to find publication includes valuable insights into the gold mines around Juneau, Alaska. The publication includes important details into the history, geology and locations of the well known gold mines and prospects in the area, including those around Windham Bay, Holkham Bay, Port Snettisham, on Grindstone and Rhine Creeks, Gold Creek, Douglas Island, Salmon Creek, Lemon Creek, Nugget Creek, from the Mendenhall River to Berners Bay, McGinnis Creek, Montana Creek, Peterson Creek, Windfall Creek, the Eagle River, Yankee Basin, Yankee Curve, Kowee Creek and elsewhere. Not only are gold placer mines included, but also hardrock gold mines. 8.5" X 11", 224 ppgs. Retail Price: $19.99

<u>Mining in the Jumbo Basin of Alaska</u> - Unavailable since 1953, this hard to find publication includes valuable insights into the mines and geology of the Jumbo Basin. The publication includes important details into the history, geology and locations of the well known gold mines and prospects in the famous Jumbo Basin Mining Region of Alaska. 72 ppgs, 9.99

<u>The Rampart Placer Gold Region of Alaska</u> - Unavailable since 1906, this hard to find publication includes valuable insights into the placer gold mines of the Rampart Mining Region. The publication includes important details into the history, geology and locations of the well known gold mines and prospects in the famous Rampart Mining Region of Alaska. 78 ppgs, 10.99

Arizona Mining Books

<u>Mines and Mining in Northern Yuma County Arizona</u> - Originally published in 1911, this important publication on Arizona Mining has not been available for over a hundred years. Included are rare insights into the gold, silver, copper and quicksilver mines of Yuma County, Arizona together with hard to find maps and photographs. Some of the mines and mining districts featured include the Planet Copper Mine, Mineral Hill, the Clara Consolidated Mine, Viati Mine, Copper Basin prospect, Bowman Mine, Quartz King, Billy Mack, Carnation, the Wardwell and Osbourne, Valensuella Copper, the Mariquita, Colonial Mine, the French American, the New York-Plomosa, Guadalupe, Lead Camp, Mudersbach Copper Camp, Yellow Bird, the Arizona Northern (Salome Strike), Bonanza (Harqua Hala), Golden Eagle, Hercules, Socorro and others. 8.5" X 11", 144 ppgs. Retail Price: $11.99

<u>The Aravaipa and Stanley Mining Districts of Graham County Arizona</u> - Originally published in 1925, this important publication on Arizona Mining has not been available for nearly ninety years. Included are rare insights into the gold and silver mines of these two important mining districts, together with hard to find maps. 8.5" X 11", 140 ppgs. Retail Price: $11.99

Gold in the Gold Basin and Lost Basin Mining Districts of Mohave County, Arizona - This volume contains rare insights into the geology and gold mineralization of the Gold Basin and Lost Basin Mining Districts of Mohave County, Arizona that will be of benefit to miners and prospectors. Also included is a significant body of information on the gold mines and prospects of this portion of Arizona. This volume is lavishly illustrated with rare photos and mining maps. 8.5" X 11", 188 ppgs. Retail Price: $19.99

Mines of the Jerome and Bradshaw Mountains of Arizona - This important publication on Arizona Mining has not been available for ninety years. This volume contains rare insights into the geology and ore deposits of the Jerome and Bradshaw Mountains of Arizona that will be of benefit to miners and prospectors who work those areas. Included is a significant body of information on the mines and prospects of the Verde, Black Hills, Cherry Creek, Prescott, Walker, Groom Creek, Hassayampa, Bigbug, Turkey Creek, Agua Fria, Black Canyon, Peck, Tiger, Pine Grove, Bradshaw, Tintop, Humbug and Castle Creek Mining Districts. This volume is lavishly illustrated with rare photos and mining maps. 8.5" X 11", 218 ppgs. Retail Price: $19.99

The Ajo Mining District of Pima County Arizona - This important publication on Arizona Mining has not been available for nearly seventy years. This volume contains rare insights into the geology and mineralization of the Ajo Mining District in Pima County, Arizona and in particular the famous New Cornelia Mine. 8.5" X 11", 126 ppgs. Retail Price: $11.99

Mining in the Santa Rita and Patagonia Mountains of Arizona - Originally published in 1915, this important publication on Arizona Mining has not been available for nearly a century. Included are rare insights into hundreds of gold, silver, copper and other mines in this famous Arizona mining area. Details include the locations, geology, history, production and other facts of the mines of this region. 8.5" X 11", 394 ppgs. Retail Price: $24.99

Mining in the Bisbee Quadrangle of Arizona - Originally published in 1906, this important publication on Arizona Mining has not been available for nearly a century. Included are rare insights into hundreds of gold, silver, copper and other mines in this famous Arizona mining area. Details include the locations, geology, history, production and other facts of the mines of this important mining region. 8.5" X 11", 188 ppgs. Retail Price: $14.99

Placer Gold Mining in Arizona - Unavailable since 1922, this hard to find publication includes valuable insights into the placer gold mines of the Arizona. Originally released as "Placer Gold of Arizona", despite its small size, this publication includes important details into the history, geology and locations of the well known placer gold mines and prospects in the State of Arizona. 48 ppgs, 8.99

Gold and Copper Mining near Payson, Arizona - Written in 1915, this hard to find publication includes valuable insights into the gold and copper mining industry of Arizona. Highlighted here are the gold and copper mines near Payson, Arizona. 68 ppgs, 8.99

Lode Gold Mining in Arizona - Unavailable since 1934, this hard to find publication, originally released as "Arizona Lode Gold Mines and Gold Mining" includes valuable insights into the gold mining industry of Arizona. Included are valuable insights into over 150 hardrock gold mines in over 30 different mining districts in Arizona. 278 ppgs, 21.99

Mining in the Dragoon Quadrangle of Cochise County, Arizona - Unavailable since 1964, this hard to find publication includes valuable insights into the mines of the Dragoon Quadrangle Mining Region. The publication includes important details into the history, geology and locations of the well known mines and prospects in this famous mining region of Arizona. 224 ppgs., 19.99

Directory of Operating Mines in Arizona in 1915 - Unavailable since 1916, this hard to find publication includes valuable insights into the mines of Arizona. This small publication includes a complete list of the mines that were operating in the State of Arizona during 1915 and includes details such as general location, owners and some basic facts about each mining operation.52 ppgs. 8.99

Arizona Ore Deposits - Unavailable since 1938, this hard to find publication includes valuable insights into some ore deposits of Arizona. Included are valuable insights into the formation and characteristics of valuable ore deposits in the Jerome, Miami, Inspiration, Clifton, Morenci, Ray, Ajo, Eureka, Tombstone and Magma mining districts. Included are details into some of the major gold, silver and copper mines of these important Arizona mining areas. 160 ppgs, 14.99

Montana Mining Books

A History of Butte Montana: The World's Greatest Mining Camp - First published in 1900 by H.C. Freeman, this important publication sheds a bright light on one of the most important mining areas in the history of The West. Together with his insights, as well as rare photographs of the periods, Harry Freeman describes Butte and its vicinity from its early beginnings, right up to its flush years when copper flowed from its mines like a river. At the time of publication, Butte, Montana was known worldwide as "The Richest Mining Spot On Earth" and produced not only vast amounts of copper, but also silver, gold and other metals from its mines. Freeman illustrates, with great detail, the most important mines in the vicinity of Butte, providing rare details on their owners, their history and most importantly, how the mines operated and how their treasures were extracted. Of particular interest are the dozens of rare photographs that depict mines such as the famous Anaconda, the Silver Bow, the Smoke House, Moose, Paulin, Buffalo, Little Minah, the Mountain Consolidated, West Greyrock, Cora, the Green Mountain, Diamond, Bell, Parnell, the Neversweat, Nipper, Original and many others. **8.5" X 11", 142 ppgs. Retail Price: $12.99**

The Butte Mining District of Montana - This important publication on Montana Mining has not been available for over a century. Included are rare insights into the gold, copper and silver mines of Butte, Montana together with hard to find maps and photographs. Some of the topics include the early history of gold, silver and copper mining in the Butte area, insight into the geology of its mining areas, the local distribution of gold, silver and copper ores, as well their composition and how to identify them. Also included are detailed facts about the mines in the Butte Mining District, including the famous Anaconda Mine, Gagnon, Parrot, Blue Vein, Moscow, Poulin, Stella, Buffalo, Green Mountain, Wake Up Jim, the Diamond-Bell Group, Mountain Consolidated, East Greyrock, West Greyrock, Snowball, Corra, Speculator, Adirondack, Miners Union, the Jessie-Edith May Group, Otisco, Iduna, Colorado, Lizzie, Cambers, Anderson, Hesperus, Preferencia and dozens of others. **8.5" X 11", 298 ppgs. Retail Price: $24.99**

Mines of the Helena Mining Region of Montana - This important publication on Montana Mining has not been available for over a century. Included are rare insights into the gold, copper and silver mines of the vicinity of Helena, Montana, including the Marysville Mining District, Elliston Mining District, Rimini Mining District, Helena Mining District, Clancy Mining District, Wickes Mining District, Boulder and Basin Mining Districts and the Elkhorn Mining District. Some of the topics include the early history of gold, silver and copper mining in the Helena area, insight into the geology of its mining areas, the local distribution of gold, silver and copper ores, as well their composition and how to identify them. Also included are detailed facts, history, geology and locations of over one hundred gold, silver and copper mines in the area . **8.5" X 11", 162 ppgs, Retail Price: $14.99**

Mines and Geology of the Garnet Range of Montana - This important publication on Montana Mining has not been available for over a century. Included are rare insights into the gold, copper and silver mines of the vicinity of this important mining area of Montana. Some of the topics include the early history of gold, silver and copper mining in the Garnet Mountains, insight into the geology of its mining areas, the local distribution of gold, silver and copper ores, as well their composition and how to identify them. Also included are detailed facts, history, geology and locations of numerous gold, silver and copper mines in the area . **8.5" X 11", 100 ppgs, Retail Price: $11.99**

Mines and Geology of the Philipsburg Quadrangle of Montana - This important publication on Montana Mining has not been available for over a century. Included are rare insights into the gold, copper and silver mines of the vicinity of this important mining area of Montana. Some of the topics include the early history of gold, silver and copper mining in the Philipsburg Quadrangle, insight into the geology of its mining areas, the local distribution of gold, silver and copper ores, as well their composition and how to identify them. Also included are detailed facts, history, geology and locations of over one hundred gold, silver and copper mines in the area **8.5" X 11", 290 ppgs, Retail Price: $24.99**

Geology of the Marysville Mining District of Montana - Included are rare insights into the mining geology of the Marysville Mining District. Some of the topics include the early history of gold, silver and copper mining in the area, insight into the geology of its mining areas, the local distribution of gold, silver and copper ores, as well their composition and how to identify them. Also included are detailed facts, history, geology and locations of gold, silver and copper mines in the area **8.5" X 11", 198 ppgs, Retail Price: $19.99**

The Geology and Mines of Northern Idaho and North Western Montana- See listing under Idaho.

The History of Gold Dredging in Montana - Unavailable since 1916, this important publication was originally published by the Us Bureau of Mines and has been unavailable for a century. A century and more ago, giant dredging machines dug in Montana's rivers and creeks in search of illusive golden riches. First appearing in California in the 1850's, gold dredges finally reached their peak of development in Siberia and New Zealand before becoming popular again in the United States. This book offers a unique historical perspective on the gold dredges that once operated in Montana. This book on Montana mining history is lavishly illustrated with dozens of rare historic photos gold dredges that once operated in Montana, as well as hard to locate plans on how these dredges were designed. 120 ppgs., 11.99

Nevada Mining Books

The Bull Frog Mining District of Nevada - Unavailable since 1910, this publication was originally compiled by the United States Department of Interior. This volume also includes important insights into the geologic formations, faults and other aspects of economic geology in this Nevada mining district. Of particular interest are the fine details on many mines in the area, including their locations, histories, development and mineralization. Some of the mines featured include the National Bank Mine, Providence, Gibraltor, Tramps, Denver, Original Bullfrog, Gold Bar, Mayflower, Homestake-King and other mines and prospects. 8.5" X 11", 152 ppgs, Retail Price: $14.99

History of the Comstock Lode - Unavailable since 1876, this publication was originally released by John Wiley & Sons. This volume also includes important insights into the famous Comstock Lode of Nevada that represented the first major silver discovery in the United States. During its spectacular run, the Comstock produced over 192 million ounces of silver and 8.2 million ounces of gold. Not only did the Comstock result in one of the largest mining rushes in history and yield immense fortunes for its owners, but it made important contributions to the development of the State of Nevada, as well as neighboring California. Included here are important details on not only the early development and history of the Comstock, but also rare early insight into its mines, ore and its geology. 8.5" X 11", 244 ppgs, Retail Price: $19.99

The Pioche Mining District of Nevada - First published in 1932, it has been unavailable for over a century and sheds important light on the mining history of Nevada. Some of the topics include the history of mining in this district, as well as the characteristics of its mineral and ore deposits. Also included are insights into the history, production, characteristics and locations of numerous mines in the area. Some of the mines include the Combined Metals, Pioche, Ely Valley, No. 10, Poorman, Wide Awake, Alps, Prince, Virginia Louise, Half Moon, Abe Lincoln, Fairview, Bristol Silver, National, Vesuvius, Inman, Tempest, Hillside, Jackrabbit, Lucky Star, Fortuna, Mendha, Manhattan, Hamburg, Comet, Lyndon and others. 108 ppgs 10.99

The Yerington Mining District of Nevada - First published in 1932, it has been unavailable for over a century and sheds important light on the mining history of Nevada. Some of the topics include the history of mining in this district, as well as the characteristics of its mineral and ore deposits. Also included are insights into the history, production, characteristics and locations of numerous mines in the area. Some of the mines include the Bluestone, Mason Valley, Malachite, McConnell, Greenwood, Western Nevada, Ludwig, Douglas Hill, Casting Copper, Montana-Yerington, Empire, Jim Beatty, Terry and McFarland, Blue Jay and others. 92 ppgs, 10.99

The Genesis of the Ores of Tonopah Nevada - Unavailable since 1918, this hard to find publication includes valuable insights into the gold mines around Tonopah, Nevada. The publication includes important details into the geology of mines in the Tonopah Mining District of Nevada. 90 ppgs, 10.99

Mining Camps of Elko, Lander and Eureka Counties Nevada - Unavailable since 1910, this hard to find publication includes valuable insights into the mining camps of Elko, Lander and Eureka Counties, Nevada. The publication includes important details into the history of mines and mining in these three Nevada counties. 154 ppgs, 12.99

Ore Deposits of the Bullfrog Quadrangle - Unavailable since 1964 and released as "Geology of Bullfrog Quadrangle and Ore Deposits Related to Bullfrog Hills Caldera, Nye County, Nevada and Inyo County, California". The publication includes important details into the geology of mines in the Bullfrog Quadrangle of Nye County, Nevada and Inyo County, California. 52 ppgs, 9.99

Mining in Eureka County Nevada - Unavailable since 1879, this hard to find publication includes valuable insights into the early mining history off Eureka County, Nevada. The publication includes important details into the early history of the mines of Eureka County, as well as their development, production and how their ores were treated. Also included are details on the 1872 Mining Act, as well as the local rules, regulations and customs of the miners in Eureka County. 134 ppgs, 12.99

Colorado Mining Books

Ores of The Leadville Mining District - Unavailable since 1926, this publication was originally compiled by the United States Department of Interior. This volume also includes important insights into the ores and mineralization of the Leadville Mining District in Colorado. Topics include historic ore prospecting methods, local geology, insights into ore veins and stockworks, the local trend and distribution of ore channels, reverse faults, shattered rock above replacement ore bodies, mineral enrichment in oxidized and sulphide zones and more. **8.5" X 11", 66 ppgs, Retail Price: $8.99**

Mining in Colorado - Unavailable since 1926, this publication was originally compiled by the United States Department of Interior. This volume also includes important insights into the mining history of Colorado from its early beginnings in the 1850's right up to the mid 1920's. Not only is Colorado's gold mining heritage included, but also its silver, copper, lead and zinc mining industry. Each mining area is treated separately, detailing the development of Colorado's mines on a county by county basis. **8.5" X 11", 284 ppgs, Retail Price: $19.99**

Gold Mining in Gilpin County Colorado - Unavailable since 1876, this publication was originally compiled by the Register Steam Printing House of Central City, Colorado. A rare glimpse at the gold mining history and early mines of Gilpin County, Colorado from their first discovery in the 1850's up to the "flush years" of the mid 1870's. Of particular interest is the history of the discovery of gold in Gilpin County and details about the men who made those first strikes. Special focus is given to the early gold mines and first mining districts of the area, many of which are not detailed in other books on Colorado's gold mining history. **8.5" X 11", 156 ppgs, Retail Price: $12.99**

Mining in the Gold Brick Mining District of Colorado - Important insights into the history of the Gold Brick Mining District, as well as its local geography and economic geology. Also included are the histories and locations of historic mines in this important Colorado Mining District, including the Cortland, Carter, Raymond, Gold Links, Sacramento, Bassick, Sandy Hook, Chronicle, Grand Prize, Chloride, Granite Mountain, Lucille, Gray Mountain, Hilltop, Maggie Mitchell, Silver Islet, Revenue, Roosevelt, Carbonate King and others. In addition to hardrock mining, are also included are details on gold placer mining in this portion of Colorado. **8.5" X 11", 140 ppgs, Retail Price: $12.99**

Ore Deposits of the London Fault of Colorado - First published in 1941, it has been unavailable since those days and sheds important light on the mines and mineral deposits of the London Fault in Central Colorado's Alma Mining District. This publication sheds important light on the gold veins and lead-silver deposits of the Alma Mining District. Included are geologic details on the London Mine, American Mine, Havigorst Tunnel, Ophir Mine, Mosher Tunnel, London-Butte Mine, Venture Shaft, Hard-To-Beat Mine, Oliver Twist Tunnel, Sacramento Mine, Mudsill Mine, Sherwood Mine, Wagner, Barcoe Tunnel and other mines in this important mining region. 110 ppgs., 10.99

The Mines of Colorado - First published in 1867, it has been unavailable since those days and sheds important light on Colorado's early mining history. Written shortly after the events took place, this publication sheds important light on the Pike's Peak Gold Rush, the discovery of gold on Ralston Creek and Dry Creek in the 1850's, as well as details on the first wave of miners into Colorado and their trials and tribulations as they crossed the Great Plains. Also included are details on early discoveries of lode gold in the mountainous regions of Colorado, details on the early mines hardrock and placer mines, and much more. It is a veritable treasure trove on Colorado's early mining history and will be of great importance to anyone who is interested in the mining of gold or other minerals in Colorado, as well as those interested in the history of the state. 478 ppgs., 29.99

The La Plata Mining District of Colorado - Originally titled "Geology and Ore Deposits in the Vicinity of the La Plata District of Colorado" and first published in 1949, it has been unavailable since those days and sheds important light on the mines and mineral deposits of the La Plata Mining District of Colorado. 214 ppgs., 19.99

Washington Mining Books

The Republic Mining District of Washington - Unavailable since 1910, this important publication was originally published by the Washington Geologic Survey and has been unavailable for a century. Topics include the geology, rock formations and the formation of ore deposits in this important mining area of Washington State. Also included are hard to find details on the geology, history and locations of dozens of mines in the area. Some of the mines featured include the New Republic Mine, Ben Hur, Morning Glory, the South Republic Mine, Quilp, Surprise, Black Tail, Lone Pine, San Poil, Mountain Lion, Tom Thumb, Elcaliph and many others. **8.5" X 11", 94 ppgs, Retail Price: $10.99**

The Myers Creek and Nighthawk Mining Districts of Washington - Unavailable since 1911, this important publication was originally published by the Washington Geologic Survey and has been unavailable for a century. Topics include the geology, rock formations and the formation of ore deposits in these important mining areas of Washington State. Also included are hard to find details on the geology, history and locations of dozens of mines in the area. Some of the mines featured include the Grant Mine, Monterey, Nip and Tuck, Myers Creek, Number Nine, Neutral, Rainbow, Aztec, Crystal Butte, Apex, Butcher Boy, Molson, Mad River, Olentangy, Delate, Kelsey, Golden Chariot, Okanogan, Ohio, Forty-Ninth Parallel, Nighthawk, Favorite, Little Chopaka, Summit, Number One, California, Peerless, Caaba, Prize Group, Ruby, Mountain Sheep, Golden Zone, Rich Bar, Similkameen, Kimberly, Triune, Hiawatha, Trinity, Hornsilver, Maquae, Bellevue, Bullfrog, Palmer Lake, Ivanhoe, Copper World and many others. **8.5" X 11", 136 ppgs, Retail Price: $12.99**

The Blewett Mining District of Washington - Unavailable since 1911, this important publication was originally published by the Washington Geologic Survey and has been unavailable for a century. Topics include the geology, rock formations and the formation of ore deposits in this important mining area of Washington State. Also included are hard to find details on the geology, history and locations of dozens of mines in the area. Some of the mines featured include the Washington Meteor, Alta Vista, Pole Pick, Blinn, North Star, Golden Eagle, Tip Top, Wilder, Golden Guinea, Lucky Queen, Blue Bell, Prospect, Homestake, Lone Rock, Johnson, and others. **8.5" X 11", 134 ppgs, Retail Price: $12.99**

Silver Mining In Washington - Unavailable since 1955, this important publication was originally published by the Washington Geologic Survey. Featured are the hard to find locations and details pertaining to Washington's silver mines. **8.5" X 11", 180 ppgs, Retail Price: $15.99**

The Mines of Snohomish County Washington - Unavailable since 1942, this important publication was originally published by the Washington Geologic Survey and has been unavailable for seventy years. Featured are details on a large number of gold, silver, copper, lead and other metallic mineral mines. Included are the locations of each historic mine, along with information on the commodity produced. **8.5" X 11", 98 ppgs, Retail Price: $10.99**

The Mines of Chelan County Washington - Unavailable since 1943, this important publication was originally published by the Washington Geologic Survey and has been unavailable for seventy years. Featured are details on a large number of gold, silver, copper, lead and other metallic mineral mines. Included are the locations of each historic mine, along with information on the commodity. **8.5" X 11", 88 ppgs, Retail Price: $9.99**

Metal Mines of Washington - Unavailable since 1921, this important publication was originally published by the Washington Geologic Survey and has been unavailable for nearly ninety years. Widely considered a masterpiece on the Washington Mining Industry, "Metal Mines of Washington" sheds light on the important details of Washington's early mining years. Featured are details on hundreds of gold, silver, copper, lead and other metallic mineral mines. Included are hard to find details on the mineral resources of this state, as well as the locations of historic mines. Lavishly illustrated with maps and historic photos and complete with a glossary to explain any technical terms found in the text, this is one of the most important works on mining in the State of Washington. No prospector or miner should be without it if they are interested in mining in Washington. **8.5" X 11", 396 ppgs, Retail Price: $24.99**

Gem Stones In Washington - Unavailable since 1949, this important publication was originally published by the Washington Geologic Survey and has been unavailable since first published. Included are details on where to find naturally occurring gem stones in the State of Washington, including quartz crystal, amethyst, smoky quartz, milky quartz, agates, bloodstone, carnelian, chert, flint, jasper, onyx, petrified wood, opal, fire opal, hyalite and others. **8.5" X 11", 54 ppgs, Retail Price: $8.99**

The Covada Mining District of Washington - Unavailable since 1913, this important publication was originally published by the Washington Geologic Survey and has been unavailable for a century. Topics include the geology, rock formations and the formation of ore deposits in this important mining area of Washington State. Also included are hard to find details on the geology, history and locations of dozens of mines in the area. Some of the mines featured include the Admiral, Advance, Algonkian, Big Bug, Big Chief, Big Joker, Black Hawk, Black Tail, Black Thorn, Captain, Cherokee Strip, Colorado, Dan Patch, Dead Shot, Etta, Good Ore, Greasy Run, Great Scott, Idora, IXL, Jay Bird, Kentucky Bell, King Solomon, Laurel, Laura S, Little Jay, Meteor, Neglected, Northern Light, Old Nell, Plymouth Rock, Polaris, Quandary, Reserve, Shoo Fly, Silver Plume, Three Pines, Vernie, White Rose and dozens of others. **8.5" X 11", 114 ppgs, Retail Price: $10.99**

The Index Mining District of Washington - Unavailable since 1912, this important publication was originally published by the Washington Geologic Survey and has been unavailable for a century. Topics include the geology, rock formations and the formation of ore deposits in this important mining area of Washington State. Also included are hard to find details on the geology, history and locations of dozens of mines in the area. Some of the mines featured include the Sunset, Non-Pareil, Ethel Consolidated, Kittaning, Merchant, Homestead, Co-operative, Lost Creek, Uncle Sam, Calumet, Florence-Rae, Bitter Creek, Index Peacock, Gunn Peak, Helena, North Star, Buckeye. Copper Bell, Red Cross and others. 8.5" X 11", 114 ppgs, Retail Price: $11.99

Mining & Mineral Resources of Stevens County Washington - Unavailable since 1920, this important publication was originally published by the Washington Geologic Survey and has been unavailable for a century. Topics include the geology, rock formations and the formation of ore deposits in these important mining areas of Washington State. Also included are hard to find details on the geology, history and locations of hundreds of mines in the area. 8.5" X 11", 372 ppgs, Retail Price: $24.99

The Mines and Geology of the Loomis Quadrangle Okanogan County, Washington - Unavailable since 1972, this important publication was originally published by the Washington Geologic Survey and has been unavailable for a century. Topics include the geology, rock formations and the formation of ore deposits in this important mining area of Washington State. Also included are hard to find details on the geology, history and locations of dozens of gold, copper, silver and other mines in the area. 8.5" X 11", 150 ppgs, Retail Price: $12.99

The Conconully Mining District of Okanogan County Washington - Unavailable since 1973, this important publication was originally published by the Washington Geologic Survey and has been unavailable for a century. Topics include the geology, rock formations and the formation of ore deposits in this important mining area of Washington State, which also includes Salmon Creek, Blue Lake and Galena. Also included are hard to find details on the geology, mining history and locations of dozens of mines in the area. Some of the mines include Arlington, Fourth of July, Sonny Boy, First Thought, Last Chance, War Eagle-Peacock, Wheeler, Mohawk, Lone Star, Woo Loo Moo Loo, Keystone, Hughes, Plant-Callahan, Johnny Boy, Leuena, Gubser, John Arthur, Tough Nut, Homestake, Key and many others 8.5" X 11", 68 ppgs, Retail Price: $8.99

Wyoming Mining Books

Mining in the Laramie Basin of Wyoming - Unavailable since 1909, this publication was originally compiled by the United States Department of Interior. Also included are insights into the mineralization and other characteristics of this important mining region, especially in regards to coal, limestone, gypsum, bentonite clay, cement, sand, clay and copper. 8.5" X 11", 104 ppgs, Retail Price: $11.99

New Mexico Mining Books

The Mogollon Mining District of New Mexico - Unavailable since 1927, this important publication was originally published by the US Department of Interior and has been unavailable for 80 years. Topics include the geology, rock formations and the formation of ore deposits in this important mining area in New Mexico. Of particular focus is information on the history and production of the ore deposits in this area, their form and structure, vein filling, their paragenesis, origins and ore shoots, as well as oxidation and supergene enrichment. Also included are hard to find details, including the descriptions and locations of numerous gold, silver and other types of mines, including the Eureka, Pacific, South Alpine, Great Western, Enterprise, Buffalo, Mountain View, Floride, Gold Dust, Last Chance, Deadwood, Confidence, Maud S., Deep Down, Little Fanney, Trilby, Johnson, Alberta, Comet, Golden Eagle, Cooney, Queen, the Iron Crown, Eberle, Clifton, Andrew Jackson mine, Mascot and others. 8.5" X 11", 144 ppgs, Retail Price: $12.99

The Percha Mining District of Kingston New Mexico - Unavailable since 1883, this important publication was originally published by the Kingston Tribune and has been unavailable for over one hundred and thirty five years. Having been written during the earliest years of gold and silver mining in the Percha Mining District, unlike other books on the subject, this work offers the unique perspective of having actually been written while the early mining history of this area was still being made. In fact, the work was written so early in the development of this area that many of the notable mines in the Percha District were less than a few years old and were still being operated by their original discoverers with the same enthusiasm as when they were first located. Included are hard to find details on the very earliest gold and silver mines of this important mining district near Kingston in Sierra County, New Mexico. 8.5" X 11", 68 ppgs, Retail Price: $9.99

East Coast Mining Books

The Gold Fields of the Southern Appalachians - Unavailable since 1895, this important publication was originally published by the US Department of Interior and has been unavailable for nearly 120 years. Topics include the geology, rock formations and the formation of ore deposits in this important mining area of the American South. Of particular focus is information on the history and statistics of the ore deposits in this area, their form and structure and veins. Also included are details on the placer gold deposits of the region. The gold fields of the Georgian Belt, Carolinian Belt and the South Mountain Mining District of North Carolina are all treated in descriptive detail. Included are hard to find details, including the descriptions and locations of numerous gold mines in Georgia, North Carolina and elsewhere in the American South. Also included are details on the gold belts of the British Maritime Provinces and the Green Mountains. **8.5" X 11", 104 ppgs, Retail Price: $9.99**

Gold Rush Tales Series

Millions in Siskiyou County Gold - In this first volume of the "Gold Rush Tales" series, leading mining historian and editor Kerby Jackson, introduces us to the story of how millions of dollars worth of gold was discovered in Siskiyou County during the California Gold Rush. Lavishly illustrated with photos from the 19th Century, this hard to find information was first published in 1897 and sheds important light onto the gold rush era in Siskiyou County, California and the experiences of the men who dug for the gold and actually found it. **8.5" X 11", 82 ppgs, Retail Price: $9.99**

The California Rand in the Days of '49 - In this second volume of the "Gold Rush Tales" series, leading mining historian and editor Kerby Jackson, introduces us to four tales from the California Gold Rush. Lavishly illustrated with photos from the 19th Century, this hard to find information was first published in 1890's and includes the stories of "California's Rand", details about Chinese miners, how one early miner named Baker struck it rich and also the story of Alphonzo Bowers, who invented the first hydraulic gold dredge. **8.5" X 11", 54 ppgs, Retail Price: $9.99**

More Mining Books

Prospecting and Developing A Small Mine - Topics covered include the classification of varying ores, how to take a proper ore sample, the proper reduction of ore samples, alluvial sampling, how to understand geology as it is applied to prospecting and mining, prospecting procedures, methods of ore treatment, the application of drilling and blasting in a small mine and other topics that the small scale miner will find of benefit. **8.5" X 11", 112 ppgs, Retail Price: $11.99**

Timbering For Small Underground Mines - Topics covered include the selection of caps and posts, the treatment of mine timbers, how to install mine timbers, repairing damaged timbers, use of drift supports, headboards, squeeze sets, ore chute construction, mine cribbing, square set timbering methods, the use of steel and concrete sets and other topics that the small underground miner will find of benefit. This volume also includes twenty eight illustrations depicting the proper construction of mine timbering and support systems that greatly enhance the practical usability of the information contained in this small book. **8.5" X 11", 88 ppgs. Retail Price: $10.99**

Timbering and Mining - A classic mining publication on Hard Rock Mining by W.H. Storms. Unavailable since 1909, this rare publication provides an in depth look at American methods of underground mine timbering and mining methods. Topics include the selection and preservation of mine timbers, drifting and drift sets, driving in running ground, structural steel in mine workings, timbering drifts in gravel mines, timbering methods for driving shafts, positioning drill holes in shafts, timbering stations at shafts, drainage, mining large ore bodies by means of open cuts or by the "Glory Hole" system, stoping out ore in flat or low lying veins, use of the "Caving System", stoping in swelling ground, how to stope out large ore bodies, Square Set timbering on the Comstock and its modifications by California miners, the construction of ore chutes, stoping ore bodies by use of the "Block System", how to work dangerous ground, information on the "Delprat System" of stoping without mine timbers, construction and use of headframes and much more. This volume provides a reference into not only practical methods of mining and timbering that may be employed in narrow vein mining by small miners today, but also rare insights into how mines were being worked at the turn of the 19th Century. **8.5" X 11", 288 ppgs. Retail Price: $24.99**

A Study of Ore Deposits For The Practical Miner - Mining historian Kerby Jackson introduces us to a classic mining publication on ore deposits by J.P. Wallace. First published in 1908, it has been unavailable for over a century. Included are important insights into the properties of minerals and their identification, on the occurrence and origin of gold, on gold alloys, insights into gold bearing sulfides such as pyrites and arsenopyrites, on gold bearing vanadium, gold and silver tellurides, lead and mercury tellurides, on silver ores, platinum and iridium, mercury ores, copper ores, lead ores, zinc ores, iron ores, chromium ores, manganese ores, nickel ores, tin ores, tungsten ores and others. Also included are facts regarding rock forming minerals, their composition and occurrences, on igneous, sedimentary, metamorphic and intrusive rocks, as well as how they are geologically disturbed by dikes, flows and faults, as well as the effects of these geologic actions and why they are important to the miner. Written specifically with the common miner and prospector in mind, the book will help to unlock the earth's hidden wealth for you and is written in a simple and concise language that anyone can understand. **8.5" X 11", 366 ppgs. Retail Price: $24.99**

Mine Drainage - Unavailable since 1896, this rare publication provides an in depth look at American methods of underground mine drainage and mining pump systems. This volume provides a reference into not only practical methods of mining drainage that may be employed in narrow vein mining by small miners today, but also rare insights into how mines were being worked at the turn of the 19th Century. **8.5" X 11", 218 ppgs. Retail Price: $24.99**

Fire Assaying Gold, Silver and Lead Ores - Unavailable since 1907, this important publication was originally published by the Mining and Scientific Press and was designed to introduce miners and prospectors of gold, silver and lead to the art of fire assaying. Topics include the fire assaying of ores and products containing gold, silver and lead; the sampling and preparation of ore for an assay; care of the assay office, assay furnaces; crucibles and scorifiers; assay balances; metallic ores; scorification assays; cupelling; parting' crucible assays, the roasting of ores and more. This classic provides a time honored method of assaying put forward in a clear, concise and easy to understand language that will make it a benefit to even beginners. **8.5" X 11", 96 ppgs. Retail Price: $11.99**

Methods of Mine Timbering - Originally published in 1896, this important publication on mining engineering has not been available for nearly a century. Included are rare insights into historical methods of timbering structural support that were used in underground metal mines during the California that still have a practical application for the small scale hardrock miner of today. **8.5" X 11", 94 ppgs. Retail Price: $10.99**

The Enrichment of Copper Sulfide Ores - First published in 1913, it has been unavailable for over a century. Topics include the definition and types of ore enrichment, the oxidation of copper ores, the precipitation of metallic sulfides. Also included are the results of dozens of lab experiments pertaining to the enrichment of sulfide ores that will be of interest to the practical hard rock mine operator in his efforts to release the metallic bounty from his mine's ore. **8.5" X 11", 92 ppgs. Retail Price: $9.99**

A Study of Magmatic Sulfide Ores - Unavailable since 1914, this rare publication provides an in depth look at magmatic sulfide ores. Some of the topics included are the definition and classification of magmatic ores, descriptions of some magmatic sulfide ore deposits known at the time of publication including copper and nickel bearing pyrrohitic ore bodies, chalcopyrite-bornite deposits, pyritic deposits, magnetite-ileminite deposits, chromite deposits and magmatic iron ore deposits. Also included are details on how to recognize these types of ore deposits while prospecting for valuable hardrock minerals. **8.5" X 11", 138 ppgs. Retail Price: $11.99**

The Cyanide Process of Gold Recovery - Unavailable since 1894 and released under the name "The Cyanide Process: Its Practical Application and Economical Results", this rare publication provides an in depth look at the early use of cyanide leaching for gold recovery from hardrock mine ores. This volume provides a reference into the early development and use of cyanide leaching to recover gold. **8.5" X 11", 162 ppgs. Retail Price: $14.99**

California Gold Milling Practices - Unavailable since 1895 and released under the name "California Gold Practices", this rare publication provides an in depth look at early methods of milling used to reduce gold ores in California during the late 19th century. This volume provides a reference into the early development and use of milling equipment during the earliest years of the California Gold Rush up to the age of the Industrial Revolution. Much of the information still applies today and will be of use to small scale miners engaging in hardrock mining. **8.5" X 11", 104 ppgs. Retail Price: $10.99**

Leaching Gold and Silver Ores With The Plattner and Kiss Processes - Mining historian Kerby Jackson introduces us to a classic mining publication on the evaluation and examination of mines and prospects by C.H. Aaron. First published in 1881, it has been unavailable for over a century and sheds important light on the leaching of gold and silver ores with the Plattner and Kiss processes. **8.5" X 11", 204 ppgs. Retail Price: $15.99**

The Metallurgy of Lead and the Desilverization of Base Bullion - First published in 1896, it has been unavailable for over a century and sheds important light on the the recovery of silver from lead based ores. Some of the topics include the properties of lead and some of its compounds, lead ores such as galenite, anglesite, cerussite and others, the distribution of lead ores throughout the United States and the sampling and assaying of lead ores. Also covered is the metallurgical treatment of lead ores, as well as the desilverization of lead by the Pattinson Process and the Parkes Process. Hofman's text has long been considered one of the most important early works on the recovery of silver from lead based ores. 8.5″ X 11″, 452 ppgs. **Retail Price: $29.99**

Ore Sampling For Small Scale Miners - First published in 1916, it has been unavailable for over a century and sheds important light on historic methods of ore sampling in hardrock mines. Topics include how to take correct ore samples and the conditions that affect sampling, such as their subdivision and uniformity. Particular detail is given to methods of hand sampling ore bodies by grab sample, pipe sample and coning, as well as sampling by mechanical methods. Also given are insights into the screening, drying and grinding processes to achieve the most consistent sample results and much more. 8.5″ X 11″, 124 ppgs. **Retail Price: $12.99**

The Extraction of Silver, Copper and Tin from Ores - First published in 1896, it has been unavailable for over a century and sheds important light on how historic miners recovered silver, copper and tin from their mining operations. The book is split into three sections, including a discussion on the Lixiviation of Silver Ores, the mining and treatment of copper ores as practiced at Tharsis, Spain and the smelting of tin as it was practiced by metallurgists at Pulo Brani, Singapore. Also included is an overview and analysis of these historic metal recovery methods that will be of benefit to those interested in the extraction of silver, copper and tin from small mines. 8.5″ X 11″, 118 ppgs. **Retail Price: $14.99**

The Roasting of Gold and Silver Ores - First published in 1880, it has been unavailable for over a century and sheds important light on how historic miners recovered gold and silver rom their mining operations. Topics include details on the most important silver and free milling gold ores, methods of desulphurization of ores, methods of deoxidation, the chlorination of ores, methods and details on roasting gold and silver ores, notes on furnaces and more. Also included are details on numerous methods of gold and silver recovery, including the Ottokar Hofman's Process, the Patera Process, Kiss Process, Augustin Process, Ziervogel Process and others. 8.5″ X 11″, 178 ppgs. **Retail Price: $19.99**

The Examination of Mines and Prospects - First published in 1912, it has been unavailable for over a century and sheds important light on how to examine and evaluate hardrock mines, prospects and lode mining claims. Sections include Mining Examinations, Structural Geology, Structural Features of Ore Deposits, Primary Ores and their Distribution, Types of Primary Ore Deposits, Primary Ore Shoots, The Primary Alteration of Wall Rocks, Alterations by Surface Agencies, Residual Ores and their Distribution, Secondary Ores and Ore Shoots and Vein Outcrops. This hard to find information is a must for those who are interested in owning a mine or who already own a lode mining claim and wish to succeed at quartz mining. 8.5″ X 11″, 250 ppgs. **Retail Price: $19.99**

Garnets: Their Mining, Milling and Utilization - First published in 1925, it has been unavailable since those days and sheds important light on the mining, milling and utilization of garnets. Included are details on the characteristics of garnets, where they are found and how they were mined. 78 ppgs, 10.99

Gemstones and Precious Stones of North America - Leading mining historian Kerby Jackson introduces us to a classic mining publication on the gems and precious stones of the United States, Canada and mexico. First published in 1890, it has been unavailable since those days and sheds important light on the gems and precious stones that may be found in North America. Included are chapters on diamonds, corundum, sapphire, ruby, topaz, emerald, disapore, spinel, turquoise, tourmaline, garnets, beyrl, peridot, zircon, quartz crystals, feldspars, pearls and many others. Included are details on where these gems and precious stones may be found throughout North America, as well as their characteristics. 360 ppgs, 24.99

Mining Camps and Mining Districts - First released in 1885 by Charles Howard Shinn under the title "Mining Camps: A Study in American Frontier Government", this publication offers a unique look at how early gold miners established their own forms of representative government during the California Gold Rush. Drawing on the the early mining codes of mideviel German miners in the Harz Mountains, on the mining customs of the Cornish tin miners and early Spanish mining laws introduced into California, the miners established the first governments in the American West. 340 ppgs, 24.99

BLM Field Handbook for Mineral Examiners - Leading mining historian Kerby Jackson introduces us to a classic mining publication on mine evaluation. First published in 1962, this work sheds important light on the techniques of BLM Mineral Examiners to perform validity on mining claims. 132 ppgs, 10.99

<u>**Six Months In The Gold Mines During The California Gold Rush**</u> - Unavailable since 1850, this important work is a first hand account of one "49'ers" personal experience during the great California Gold Rush, shedding important light on one of the most exciting periods in the history of not only California, but also the world. Compiled from journals written between 1847 and 1849 by E. Gould Buffum, a native of New York, "Six Months In The Gold Mines During The California Gold Rush" offers a rare look into the day to day lives of the people who came to California to work in her gold mines when the state was still a great frontier. **8.5" X 11", 290 ppgs. Retail Price: $19.99**

<u>**The Discovery of Gold in Australia**</u> - **First published in 1852, it has been unavailable since those days and sheds important light on Australia's gold mining history. Included are rare communications between British agents and the British Crown when gold was first discovered in Australia in 1851. This rare text contains hard to find details on Australia's first mining camps and Britain's early attempts to provide for the orderly regulation of gold mines in that part of the world. Also of interest are hard to find extracts of articles that appeared in the early colonial newspapers that did their best to report on Australia's gold rush as it took place.**
102 ppgs, 10.99

www.ingramcontent.com/pod-product-compliance
Lightning Source LLC
Chambersburg PA
CBHW080804180526
45168CB00006B/2324